ERGEBNISSE DER PHYSIOLOGIE
BIOLOGISCHEN CHEMIE UND
EXPERIMENTELLEN PHARMAKOLOGIE

HERAUSGEGEBEN VON

K. KRAMER
GÖTTINGEN

O. KRAYER
BOSTON

E. LEHNARTZ
MÜNSTER / WESTF.

A. v. MURALT
BERN

H. H. WEBER
HEIDELBERG

BAND 52

MIT BEITRÄGEN VON

K. BÄTTIG · H. H. DALE · H. DAVSON
A. S. PAINTAL · H. E. ROSVOLD

MIT 49 ABBILDUNGEN UND 1 PORTRÄT

SPRINGER-VERLAG
BERLIN · GÖTTINGEN · HEIDELBERG
1963

ISBN 978-3-642-49604-2   ISBN 978-3-642-49896-1 (eBook)
DOI 10.1007/978-3-642-49896-1

Inhaltsverzeichnis

Berichtigung zu Seite 10 der „Ergebnisse der Physiologie", Band 51

In der 6. Zeile des 2. Absatzes muß es richtig heißen:
. . . LANG und OTTESEN arbeiteten das Material auf . . .

Otto Loewi †

By

H. H. DALE

OTTO LOEWI, born on June 3rd, 1873 in Frankfurt a. M., was the son of JAKOB LOEWI, a wine merchant in that city, by his wife, born ANNA WILLSTAEDTER. OTTO had many happy memories of his childhood, at home and at school. His schooling was of the conventional, humanistic type of the old-fashioned Gymnasium, centred on the Latin and Greek classics. He had vivid memories also of annual summer holidays, spent at his father's small country estate in the Palatinate, on a slope of the Haardt Mountains, with a pleasant house, a large garden and vineyards. There were holiday visits also to Belgium, from which he dated a lasting enthusiasm for early Flemish painting.

So far, there had been nothing to indicate the choice, for OTTO LOEWI's further educational course, or for his eventual career, of any of the natural scientific disciplines. He has himself recorded[1] that his own most natural choice would have been for the History of Art; but his parents, doubtless for good reason, wished that he should study to qualify in medicine. He accepted their decision, and duly matriculated at Strassburg, with a medical course in view. He apparently found little to attract him in the formal studies of the preclinical course, with the exception, perhaps a rather surprising one, of those in Anatomy. He was evidently much influenced by the personal characteristics of his teachers. In any case, he seems to have spent much of his time, and nearly all his enthusiasm, during his first University year, in hearing lectures on the history of architecture and on philosophical theory; with the result that he only passed his "Physikum" at the end of his fourth term. Then he spent a year at MUNICH, where his medical studies were inevitably in competition with further and abundant opportunities for deepening and widening his knowledge and enjoyment of a range of the arts drama, music, and the treasures of museums and galleries. LOEWI retained, indeed, throughout his long life, and amid all the changes in his circumstances, the desire to use every opportunity for enlarging his artistic and cultural experiences in general, as well as for making contact with new developments in a range of the natural sciences. He has himself recorded that, having by nature no special aptitude

---

[1] Perspectives in Biology and Medicine — Autumn, 1960.

for making music, or even for an understanding enjoyment of it, he deliberately acquired an appreciation of its appeal, so that it came to take a high rank in the scale of his artistic pleasures.

On the other hand, he was evidently ready, as a student, to respond to evidence of personal or scientific distinction in his teachers. On his return in 1894 to Strassburg, a new enthusiasm for his medical studies was thus aroused in him by NAUNYN's lectures and clinics; so that, for the time, be began to think of a career in clinical medicine. He had first to find a subject, however, for a graduation thesis; and, for no special reason which he could recall, he went for this to SCHMIEDEBERG, who was then gaining a wide influence as one of the early Professors of Pharmacology. The subject which SCHMIEDEBERG gave him was not likely to afford him much more than a first experience of the relatively simple techniques, employed in the pharmacological researches of those days. The opportunity, however, brought him contacts, and more lasting friendships, with men of the calibre of OSCAR MINKOWSKI, ARTHUR CUSHNY, KARL SPIRO and WALTHER STRAUB; and it seems, indeed, to have been eventually effective in determining the main course of LOEWI's career. It was SPIRO, as LOEWI recalled, who first brought to his attention FRIEDRICH MIESCHER's classical studies of the changes in the metabolism of the salmon, during its fresh-water progress up the RHINE, and thus to the possibilities of chemical methods applied to biological problems.

When he had graduated at Strassburg, LOEWI's parents arranged for him first to enjoy a holiday in Italy, with free scope for the indulgence of his artistic interests. He had evidently retained, however, the sense of a deficiency on the chemical side of his training, for the future needs of a scientific career in Medicine; so that, on returning from the Italian holiday, he took a course of chemistry in Frankfurt before returning to Strassburg, to spend a few more months there under the stimulating influence of FRANZ HOFMEISTER, whose Department was already one of the centres of the effective emergence of the then new discipline of biochemistry. And if one looks at the list of all the scientific papers which LOEWI was to publish, during the four decades of his full activity in pharmacological teaching and research, and notes how often, and how recurrently, they were concerned with problems of nutrition and metabolism, it is easy to form the impression that he might well have chosen biochemistry as the subject of his life's work, if he had earlier had the opportunity, and the impulse, to obtain a stronger educational background in chemistry, and in organic chemistry especially.

The strong influence, however, which LOEWI had experienced from NAUNYN's clinical teaching had apparently not yet lost its effect. Before he finally decided to seek his major opportunity in pharmacology, he accordingly accepted a clinical assistantship under Professor VON NOORDEN, at the City Hospital in Frankfurt. He found the experience discouraging, however, and returned after a year of it to Strassburg, where he sought the advice of Professor HOF-MEISTER, on whose recommendation he enquired whether HANS HORST MEYER, then Professor of Pharmacology at Marburg a. d. Lahn, had a suitable vacancy in his Department. Thus he was to become, in 1898, Assistant in MEYER's department, and was to remain with him as Privat-Dozent and Associate Professor, for eleven years in all — seven years in Marburg, and four in Vienna, when MEYER was called to the Chair of Pharmacology there. During these years he acquired a deepening personal devotion to MEYER and a high regard for his scientific distinction and his wisdom; and their friendship continued after LOEWI himself, in 1909, became Professor of Pharmacology in Graz, where he was to remain in full activity until he was imprisoned and exiled by the Nazi invaders, in 1938.

Mention has been made of the biochemical and metabolic interest shown by a number of LOEWI's researches; and this was already obvious in his earlier publications from the Marburg laboratory. Already in 1898 he was describing a urea-forming enzyme in the liver, and in 1900 came several publications on the metabolism of nuclein. Even those which might be classified as pharmacological had apparently a biochemical aspect, as when phloridzin was given to excite a diabetes, and the effect of camphor on the condition so produced was studied. In 1901 he was publishing experiments giving a negative answer, to the question whether the body could form sugar from fat; and then, in 1902, came the publication of experimental results, for which a claim might well be made, as the most important of any which LOEWI achieved, until his famous demonstration, in 1921, of chemical neuro-effector transmission. In the 1902 publication LOEWI described nutritional experiments in which he had succeeded, where all others had failed till then, in maintaining dogs in nitrogenous equilibrium, on a diet in which the sole source of the nitrogen was a mixture of amino-acids, obtained by the complete tryptic digestion of the proteins of natural organs. LOEWI was in England for some months in that year, and was able to discuss these findings with the late Dr. (later Sir) FREDERICK GOWLAND HOPKINS. And I well remember a conversation in which HOPKINS spoke to me with enthusiasm, of the importance to nutritional science of this discovery; and it might, indeed, be regarded as an important step towards the experiments with a diet of chemically purified nutrients, which were later to lead HOPKINS himself to his own historic contribution to the discovery of the "vitamins".

It was in 1902 that LOEWI paid this, his first visit to England. He was anxious to improve his knowledge of the techniques of mammalian physiology, and judged that, at that time, he could learn more to his purpose in England than in Germany. With the ready agreement of the Professor of Physiology in University College, London, ERNEST H. STARLING, he made that Department the chief centre of his visit, during a stay of some months. STARLING's laboratory was full of interest at the time, on account of his then recent discovery of secretin with his equally famous brother-in-law and frequent collaborator, WILLIAM M. BAYLISS. Others were there, exploring the possibilities of different developments from this discovery — myself among them, working with the support of one of the very few post-graduate research studentships which were then available. I thus met LOEWI for the first time, and we laid the foundations of a personal friendship, and a sharing of research interests, which was to last for nearly 60 years. He had not planned to do any research of his own on this short visit. He wished to learn the methods then in use in British laboratories, to make personal contacts with British physiologists, and to learn something of the problems on which they were engaged. Also he wished to learn enough English for ordinary scientific and social purposes — not, as he said, to speak it accurately, but to speak it fast. The legend of some of the impromptu translations, which he thus achieved, had a long survival in British physiological circles, at Oxford and Cambridge also, to which he made shorter visits from his London base, and in the Physiological Society. The latter was then a relatively small and intimate community, and LOEWI found the friendly informality of its meetings and discussions much to his liking.

A special mention should be made of the short stay which he made in Cambridge, where again he found a Physiological Department full of research and ideas. Mention has already been made of his meeting with HOPKINS; and he would have met WALTER FLETCHER also, who was then working with Hopkins. He would also have met JOSEPH BARCROFT, who was later to succeed LANGLEY as Professor of Physiology. Of more eventual significance, however, was his meeting at that time with Cambridge workers on the functions of the involuntary or autonomic nervous system, on the separate and contrasted functions of its sympathetic and parasympathetic divisions, and on the closely mimetic reproduction of their respective effects by adrenaline and by muscarine. In this connexion he would have met the veteran W. H. GASKELL, no longer active in this field, J. N. LANGLEY and H. K. ANDERSON, and, more especially, T. R. ELLIOTT, who was then still a post-graduation research student, but was already engaged on his remarkably brilliant and inclusive survey of the closely sympathomimetic actions of adrenaline. There was certainly a good deal of speculation among the Cambridge group then, with regard to the meaning of this correspondence, which survived the degeneration of the nerves. I do not think that ELLIOTT would then have mentioned the conception of adrenaline as the chemical transmitter of sympathetic effects; he put this forward two years later, and thus made the first published suggestion of such a specific, chemical transmission. LOEWI, on his return to London from Cambridge, made a special mention to me of the impression which he had formed of ELLIOTT, as a research worker of brilliant potentialities. And it would appear that, when he returned to Marburg, LOEWI's mind was still busy, perhaps subconsciously, with the meaning of these neuromimetic actions, which had been so much "in the air" at Cambridge during his visit. For it is on record that, in the following year, 1903, when walking with WALTER M. FLETCHER, whom he had also met at Cambridge, and who was then on a visit to HANS MEYER's Department in Marburg, LOEWI suddenly exclaimed: "FLETCHER! Perhaps the vagus impulses inhibit the heart's action by liberating muscarine at the nerve endings!" The suggestion was evidently made quite casually. LOEWI himself promptly forgot it, and retained no conscious memory of it. It was FLETCHER who recalled it, after LOEWI's publication of his famous demonstration, in 1921. And, by that time, it was no longer necessary to think of a substance so improbable as the stable, poisonous alkaloid, muscarine, to act as the vagus-transmitter, the real existence of which had been so clearly established by LOEWI's practical demonstration.

There can be no doubt that LOEWI's high reputation as a discoverer, among scientists in general, has been mainly due to a recognition of the far-reaching importance of these experiments on chemical, neuro-effector transmission; and I must reserve a separate section of this Memoir for further details of this crowning experimental success of LOEWI's career. In the smaller circle, however, of those who are interested in the general development of the medical sciences, and of pharmacology in particular, during the first half, and especially during the first four decades, of the present century, it will certainly be felt that the fame of his greatest single achievement should not be allowed to overshadow unduly the importance of LOEWI's solid and widely ranging contributions to the advancement of knowledge, in several different sections of this pharmacological field — some of them bordering on, or even overlapping with, those of biochemistry. The mere list of his published researches during the period of his full activity from 1898 to 1938 — many of them carried out in co-operation with a succession of his departmental colleagues, or with visiting physiologists or pharmacologists from other centres and other countries — provides abundant evidence of the high rank as an investigator, and as a promoter of researches by others, to which LOEWI would in any case have been entitled, even if he had never received the mysterious impulse, to undertake the experiments which were to make his name most widely famous. A glance at the record will show that there were several subjects which, at intervals of some years, seemed to acquire a renewed interest for him, with the resulting appearance of successive groups of communications, in association with different collaborators. Such were the physiology and pharmacology of renal function, including the effects upon it of digitalis; carbohydrate metabolism, including the diabetes produced by phloridzin and the effects of other drugs upon it, and, later, after the discovery of insulin, the mode of the action of that hormone; and, again, the part played by inorganic ions, and especially by calcium ions, in the actions of drugs of the digitalis series on the functions of the heart, and in other drug actions. I have already mentioned his early and very important demonstration of the body's ability to build its proteins from their constituent amino-acids. And there were several other investigations, pharmacological in aim and method, which might have served to quicken LOEWI's interest, during the period before 1921, to the possibility of a chemical, or, as he preferred to call it, a neurohumoral transmission of nervous effects. Such were the opportunity which he had, with HANS MEYER, to test some of the substances obtained in industrial research on the way to the synthesis of adrenaline, and including "arterenol" (i.e. *nor*-adrenaline); his discovery, with FRÖHLICH, of the increased sensitiveness to the specific effects of adrenaline produced by cocaine; and, with MANSFELD, of the action of physostigmine, in specifically increasing the responses of effector organs to the stimulation of parasympathetic nerves. From LOEWI's own reminiscences, however, it seems clear that, at the time when these observations were made, he had retained none of the interest, of which he had given momentary evidence some years earlier, in the chemical transmission possibility.

Concerning the range of LOEWI's other researches, it would, perhaps, be proper to regard them as having made their most important contribution to the classical period of pharmacology, when it still had relatively small and intermittent contacts with the traditional empiricism of therapeutics. LOEWI's full activities in research, cut short by some years of their normal term, would not, I think, in any case, have extended far into the modern period, with its revolutionary transformation of medicinal therapeutics itself into an experimental science, progressively employing the resources made available by advances in endocrinology, in immunology, in the chemotherapy of infections, in the synthesis of specifically active remedies for symptoms, and, indeed, in widening ranges of physiology, biochemistry, biophysics, and even of more fundamental disciplines. By the time that this great change was gathering speed, LOEWI was resettled, under the conditions of a new and hospitable country, offering opportunity, in the happy evening of his life, to gather memories from the years of his more active interests and achievements in research, and to expound the philosophy which these had taught him.

Loewi's proof of neurohumoral (chemical) transmission

It is beyond doubt that the publications by LOEWI, beginning in 1921, of the direct evidence for neuro-effector transmission by chemical agents, which he alone and, later, with his colleagues, was able to obtain and to extend by experiments on the isolated hearts of frogs, changed the whole status and prospect of that conception. Till then, since ELLIOTT first put it forward tentatively, in 1904, to explain the closely sympathomimetic actions of adrenaline, it had remained, at best, an ingenious and, for a few physiologists, an attractive hypothesis. I had myself retained some special interest in it from my own early association with ELLIOTT in experiments bearing on his idea; and this interest had been revived for me some ten years later, by experiments which I published in 1914, showing that acetylcholine had a potent and rapidly evanescent action, reproducing the peripheral effects of parasympathetic nerve impulses, at least as closely as adrenaline reproduced those of sympathetic nerves. For me, however, as for everybody else who had thought about it, the suggestion had seemed still to be experimentally intangible; and, so long as it could not be verified by direct experiment, it must remain only a speculative, and, to that extent, a practically unfruitful hypothesis. OTTO LOEWI himself seems, until 1920, to have regarded it as even more negligible; having apparently just glanced at it in 1903, he had forgotten all about it, and, with the majority of physiologists and pharmacologists, had taken no further conscious interest in it. It is, of course, only the more remarkable that, on the eve of Easter Sunday in 1920, the suggestion of an experimentally simple and straightforward method, for putting it to a practical test, should have come to him in a dream; and it was still more remarkable that, when the note which he made on waking had proved illegible on the following morning, and his memory blank even with regard to its purport, he should have found himself waking again, on the following night, from the same insistent dream, so as to be given another chance to follow its suggestion. According to his own latest account of what followed, which we must accept as authoritative, he rose immediately from his bed at 3 a.m., and hurried to his laboratory to carry out the experiment which his dream had again suggested, avoiding thus any risk that the memory of it might fade again, before he had been able to give it a trial. In his description of the whole incident[1], LOEWI records his own belief that, if such an experiment had been suggested to him when he was fully awake, he would have rejected the idea as absurd; on the ground that, even if a chemical transmission really existed, it would be impossible to suppose that the transmitting agents would be liberated in sufficient excess, to be found, in the fluid filling the frog's heart, in concentrations sufficient to transfer the effects to another heart. Whatever might have happened under different circumstances, the fact of importance is that LOEWI was not thus inhibited from trying the experiment, and that it was immediately and completely successful. Stimulation of the vagus and sympathetic nerves caused the liberation of an inhibitory and an accelerator substance, by which the respective nervous effects could be reproduced in another heart. Atropine and ergotoxine prevented the respective effects of the transmitters thus liberated, but did not interfere with their liberation.

I was inclined to rally LOEWI on what seemed to me his excessive caution, at that early stage, in referring to the transmitters, which he had demonstrated, simply as "Vagusstoff" and "Acceleransstoff", and avoiding any mention of

---

[1] Autobiographic Sketch, loc. cit. 1960.

their suggestive similarities to acetylcholine and adrenaline respectively. The matter of real importance, however, was that the further experiments, as reported in some 19 publications in all, on these phenomena, by LOEWI and his colleagues, in the period from 1921 to 1938, were largely directed to demonstrating properties of the vagus-transmitter, so closely corresponding to those of acetylcholine, as practically to establish its identity. And of very special importance, for further developments, was the evidence which they produced, that the remarkable evanescence of the action of the vagus-transmitter, as of acetylcholine, was due to its rapid destruction by a specific cholinesterase; and, further, that the long known action of physostigmine (eserine), in specifically intensifying responses to the stimulation of parasympathetic nerves, was due to its antagonism to the action of this cholinesterase, and to the consequently longer survival of the liberated acetylcholine, and the corresponding persistence of its effects.

LOEWI's first experiments in this series, published in 1921, had already changed the status of chemical, neurohumoral transmission, from that of an interesting, but apparently intangible hypothesis, to that of an experimentally demonstrable phenomenon, at least for the peripheral, neuroeffector transmission, from autonomic nerve-endings in the amphibia. I think, however, that it should also be emphasized, that it was the later experiments of the same series, made by LOEWI with a succession of his colleagues, and dealing with the identity of the vagus transmitter, with cholinesterase and the protective effect of eserine, which opened the way, first to further developments by a number of investigators, extending the evidence for neuro-effector transmission from peripheral, autonomic nerve endings, to the warm-blooded, mammalian types; and then, yet further, to investigations, primarily by workers in my own research department, of the possible significance, for its transmitter function, of the other, "nicotine-like" actions of acetylcholine, on ganglion cells and voluntary motor end-plates. With regard to the likelihood of this last extension of the transmitter function of acetylcholine, LOEWI himself shared the sceptical attitude of some other distinguished physiologists; and he was even, as it seemed to me, incautious enough, to make a public profession of his incredulity, in a lecture. But he readily accepted, of course, when it came, the experimental evidence for these additional cholinergic phenomena; one of the essential conditions for the production of this evidence having, in fact, been the protective action of eserine, which his own experiments had revealed. And thus, in effect, LOEWI's trial in 1920, at the bidding of an impulse so strangely engendered by a dream, of so simple and direct a method for testing the chemical transmission hypothesis, and his publication in 1921 of the unquestionable evidence for it, must in fact be regarded as having provided the starting point for new research enterprises in the whole of this field, extending now to the evidence for chemical transmission, not only at all the junctions of the peripheral nervous system, but more recently, largely through the initiative of Sir JOHN ECCLES and his team collaborators, at those of the formidable synaptic complexities of the mammalian central nervous system.

The appropriateness of the award of a Nobel Prize to OTTO LOEWI, in 1936, for his researches in this field, was widely recognized. To me it was a matter for special pride and satisfaction, to be associated with him in this award; and, when we met again in Stockholm for the impressive ceremony, we both welcomed this as a new link in our scientific and personal friendship.

Personal and Family Life. Exile and Resettlement

Reference has already been made to OTTO LOEWI's early interests, retained throughout his long life, in a wide range of the arts. He seems to have found Marburg rather restricted in such cultural opportunities; but these were lavishly at his disposal, of course, when he moved with HANS MEYER to VIENNA. While there he formed the habit of spending an annual holiday in Switzerland, and formed a special affection for the Engadine, which his memory always retained; and this feeling was further confirmed when at Pontresina, in 1907, he met Dr. GUIDO GOLDSCHMIDT, then Professor of Chemistry in Prague and later in Vienna, with his wife and their daughter GUIDA, who became OTTO LOEWI's wife in 1908, and was able to share with him all the details of their removal, in 1909, to their new and attractive home in Graz. There they were to be happily established for the next 29 years — the major part of LOEWI's scientific career — and to bring up their family of three sons and one daughter. GUIDA LOEWI, devoted wife and constant companion to her husband and her children, mistress of a charming household and perfect hostess to their many friends, was to share in all her husband's interests and experiences, happy and successful, tragic and adventurous, until she died with a tragic suddenness in New York, in 1958, fifty years after their marriage, at a time when they had been about to leave together, on a first visit to Europe since their exile.

In every way, the tenure of the Chair of Pharmacology at Graz had seemed to give LOEWI, and his family, the kind of opportunities which they could most desire — admirable conditions for teaching and research in his chosen range of the sciences, and therewith the social life and friendships and the cultural privileges which they could all enjoy. I was particularly impressed by the fact that even the first world war, beginning 5 years after he became Professor in Graz, had appeared to leave LOEWI free to continue his academic researches, practically unhindered by demands to meet war-time emergencies; by contrast with the experience of his envious colleagues, myself among them, in other belligerent countries.

By MARCH 1938, LOEWI might presumably have begun to think of plans for his normal retirement, a few years later. With the Nazi occupation of Austria, however, he and his two younger sons found themselves suddenly arrested and thrown into prison, together with a large number of other Jewish residents in Graz. After two months of this deprivation and ill usage, LOEWI was released; and in September he obtained permission to leave Austria, but only at the cost of signing an instruction to a Swedish bank, to transfer property, which was there being held for him, to a German bank under State control. A cable informed me that he was flying to me in England; and, in welcoming him, we were relieved to find that his arrival had not coincided with the outbreak of a war, which had then appeared to be imminent, but, in the event, was postponed till the following year. After staying with us for a few weeks, during which he was able to make and to renew English contacts, he received, and accepted, a welcome invitation from the Belgian Fondation Franqui, and left us for Brussels, where he was installed as a Research Professor, in Professor HEYMANS's Department of Pharmacology. Being in England again, however, in the next year, for a holiday with friends in the country, he was caught by the outbreak of the second world war, and could not return to Belgium. He found harbourage, however, and renewed opportunity for some months, in the late Professor GUNN's Department of Pharmacology at Oxford. And then, in the middle of 1940, he was finally able to accept an invitation from the Medical School of the New York University, to become a Research Professor of Pharmacology in Professor GEORGE B. WALLACE's Department. LOEWI arrived in New York on June 1st 1940, two days before he became 67 years of age, and was fortunate thus to resettle himself in the further and final stage of his career; for the generosity, which had provided him with this attractive research appointment, was extended to enable him to hold it to the end of his life. It was not to be expected that, at that stage of his long career, he would be able to resume his full activity in research. He was concerned mostly to tidy a number of "loose ends"; but his daily visits to Professor WALLACE's Department were warmly welcomed, for the inspiration which contacts with his ripe and varied experience, and with his still enthusiastic interest, gave to the younger research workers and to the students.

LOEWI's wife and the members of their family were all able to join him in New York in 1941, where they readily and gratefully adapted themselves to the new conditions and associations. He became an American citizen in 1946, and he soon acquired, thus late in life, so good and natural a style, as well as so ready a fluency, in the English language, that he was still for a number of years in wide demand, as an effective and eloquent lecturer in his own extensive range. His tireless zest for knowledge, his eager brilliance in discussion and reminiscence, made him widely known, and brought him many friends during this long and happy evening of his career, which was to last for more than 21 years. He and his wife soon formed the habit of going for the hot months of each summer to Woods Hole, on the coast of Massachusetts, where the world famous Research Station for Marine Biology gave them opportunities for making contact with additional aspects of scientific research, for forming new scientific friendships, and for renewing others already established. Loewi had already been suffering for some time from asthma, and from other physical troubles, which caused some concern to his friends. And this was accentuated when, not long after the sad death of Mrs. LOEWI, he had a fall which fractured his pelvis, and left him with a severe and lasting physical disablement. When I was able to visit him, however, in New York, in the autumn of 1959, his brain had evidently retained all its familiar activity and alertness, and he was still as eager for reminiscence, for argument and humorous comment, spiced with quotations from his favourite Goethe, and for new mental enterprise. In each of his remaining summers there were friends ready to give him careful transport to Woods Hole, where he continued fully to enjoy the conditions and the company. In the last letter which I had from him, dated from Woods Hole on July 2nd, 1961, he wrote: — "I love this place more than any other in the world known to me, except the Engadine in Switzerland." It was good to think of him thus, eased and refreshed by the Atlantic breezes, where, as recently described by friends in the U.S.A.[1], "he became a legend to the students of the Marine Biological Laboratories" — their "Uncle Otto" in fact; and where "On summer afternoons and evenings he was usually seen sitting in front of the house where he lived, at his feet on the lawn a small knot of people listening to an ever animated discourse". OTTO LOEWI, indeed, enjoyed good conversation as he enjoyed all the good things, material and spiritual, which life had to offer; and he was himself an irrepressible talker, though never a wearisome one. His family realised that, when the end came, it came in a form which he might himself have desired. On the morning of December 25th, Christmas Day, 1961, he suddenly stopped speaking, in the middle of a sentence; and it was found that his life had flickered out.

On August 17th, 1962, Dr. PHILIP B. ARMSTRONG, the Director of the Marine Biological Laboratory, in conjunction with Professor STEPHEN KUFFLER of the Harvard University Medical School and other friends of OTTO LOEWI, arranged for the reinterment of his ashes at Woods Hole, where he had been so happy in so many summers, and for a commemorative celebration in connexion therewith. To my deep regret I was prevented, by a minor accident, from accepting a generous invitation to be present on this occasion and to take part in the proceedings. My friend and former colleague Sir G. LINDOR BROWN, however, now a Secretary of the Royal Society, was fortunately able to be present, and to represent OTTO LOEWI's many British friends and admirers.

---

[1] KRAYER, DAVIS and KUFFLER: The Pharmacologist, pp. 47—49. Spring 1962.

It was mostly in his later years, when he had become a citizen of the U.S.A., that OTTO LOEWI received honorary degrees from many universities, and was honoured by many Scientific Societies. He thus became Honorary Sc. D. of New York and Yale, Honorary Ph. D. and M. D. of Graz, Honorary M. D. of Frankfurt. He was made a Foreign Member of the Royal Societies of London and of Edinburgh, of the Accademia dei Lincei of Rome, of the Bavarian and Austrian Academies of Science; an Honorary Member of the Academies of Medicine of New York and of Belgium, of the American Society of Pharmacology, and of the British and German Societies of Physiology and of Pharmacology.

Published Works of Otto Loewi

Books

Untersuchungen über den Nucleinstoffwechsel. Habil.-Schr., Marburg 1900.

Arzneimittel und Gifte in ihrem Einfluß auf den Stoffwechsel. In: VON NOORDEN, Handbuch der Pathologie des Stoffwechsels, 2. Aufl., Bd. 2, S. 663—831. Berlin 1907.

Also transl.: Drugs and poisons: their influence on metabolism. In: VON NOORDEN, Metabolism and practical medicine, English issue, vol. 3, pp. 1061—1198. London 1907.

Allgemeines über Beziehungen zwischen Nervensystem und vegetativen Funktionen. In: LÜDKE-SCHLAYER, Lehrbuch der pathologischen Physiologie, pp. 258—268. Leipzig 1922.

Über postganglionäre Gefäßwirkungen von Nicotin und Acetylcholin. Arch. int. Physiol., Vol. jubilaire pour J. Demoor. 1937, pp. 349—356.

Vistas in pharmacology. Symposium on the next half century in medicine. Sterling-Winthrop Research Institute, Rensselaer, N.Y., 1950, pp. 13—24.

On the mechanism of drug action. In: E. S. G. BARRON (ed.), Modern trends in physiology and biochemistry, pp. 405—425. New York 1952.

From the workshop of discoveries. Porter Lectures, Ser. 19. (1) Reflections on the study of medicine. (2) From the workshop of discoveries. (3) Problems in the field of adrenal function. Lawrence, Kan.: Univ. Kansas Press 1955.

Periodicals

1896

Zur quantitativen Wirkung von Blausäure, Arsen und Phosphor auf das isolierte Froschherz. Naunyn-Schmiedeberg's Arch. exp. Path. Pharmak. **38**, 127—138 (1896).

1898

Über das harnstoffbildende Ferment der Leber. Hoppe-Seylers Z. physiol. Chem. **25**, 511—522 (1898).

1900

Beiträge zur Kenntnis des Nucleinstoffwechsels. I. Mitt. Naunyn-Schmiedeberg's Arch. exp. Path. Pharmak. **44**, 1—23 (1900).

Zur Kenntnis des Nucleinstoffwechsels. S.-B. Ges. Naturw. Marburg (1900), 1901, S. 89—92.

1901

Pharmacodynamische Untersuchungen über Anagyrin. Arch. int. Pharmacodyn. **8**, 65—75 (1901).

Zur Lehre von der Fettresorption. S.-B. Ges. Naturw. Marburg 1901, S. 90.

Untersuchungen über den Nucleinstoffwechsel. II. Mitt. Naunyn-Schmiedeberg's Arch. exp. Path. Pharmak. **45**, 157—185 (1900/01).

Über die Stellung der Purinkörper im menschlichen Stoffwechsel. Bemerkungen zu der gleichnamigen Untersuchung von BURIAN und SCHUR. Pflügers Arch. ges. Physiol. **88**, 296—298 (1901).

Zur Frage nach der Bildung von Zucker aus Fett. Naunyn-Schmiedeberg's Arch. exp. Path. Pharmak. **47**, 68—76 (1901/02).

Über den Einfluß des Camphers auf die Größe der Zuckerausscheidung im Phlorhizindiabetes. Naunyn-Schmiedeberg's Arch. exp. Path. Pharmak. **47**, 56—67 (1901/02).

Zur Kenntnis des Phlorhizindiabetes. Naunyn-Schmiedeberg's Arch. exp. Path. Pharmak. **47**, 48—55 (1901/02).

1902

Untersuchungen zur Physiologie und Pharmakologie der Nierenfunction. Naunyn-Schmiedeberg's Arch. exp. Path. Pharmak. **48**, 410—438 (1902).

Über Eiweißsynthese im Thierkörper. Naunyn-Schmiedeberg's Arch. exp. Path. Pharmak. **48**, 303—330 (1902). Also: Zbl. Physiol. **15**, 590—591 (1901/02).

1903

Untersuchungen zur Physiologie und Pharmakologie der Nierenfunktion. II. Mitt. Über das Wesen der Phlorhizindiurese. Naunyn-Schmiedeberg's Arch. exp. Path. Pharmak. **50**, 326—331 (1903).

1904

Pharmakologie des Wärmehaushalts. Ergebn. Physiol. **3**, 332—372 (1904).

Zur Kenntnis der Tetanusvergiftung. S.-B. Ges. Naturw. Marburg (1904), 1905 S. 11—17.

Über das Wesen der Coffeindiurese. S.-B. Ges. Naturw. Marburg (1904), 1905, S. 76—79.

Zur Physiologie und Pharmacologie der Vasodilatatorenreizung. S.-B. Ges. Naturw. Marburg (1904), 1905, S. 79—85.

Über den Diastasengehalt verschiedener Blutsera. S.-B. Ges. Naturw. Marburg (1904), 1905, S. 100—102.

1905

(With W. M. FLETCHER and V. E. HENDERSON) Untersuchungen zur Physiologie und Pharmakologie der Nierenfunction. III. Mitt. Über den Mechanismus der Coffeindiurese. Naunyn-Schmiedeberg's Arch. exp. Path. Pharmak. **53**, 15—32 (1905).

(With N. H. ALCOCK) Untersuchungen zur Physiologie und Pharmakologie der Nierenfunction. IV. Mitt. Über den Mechanismus der Salzdiurese. Naunyn-Schmiedeberg's Arch. exp. Path. Pharmak. **53**, 33—48 (1905).

(With V. E. HENDERSON) Untersuchungen zur Physiologie und Pharmakologie der Nierenfunction. V. Mitt. Über den Mechanismus der Harnstoffdiurese. Naunyn-Schmiedeberg's Arch. exp. Path. Pharmak. **53**, 49—55 (1905).

(With V. E. HENDERSON) Über die Wirkung der Vasodilatatorenreizung. Naunyn-Schmiedeberg's Arch. exp. Path. Pharmak. **53**, 56—61 (1905).

(With V. E. HENDERSON) Über den Einfluß von Pilocarpin und Atropin auf die Durchblutung der Unterkieferspeicheldrüse. Ein Beitrag zur Frage nach den Beziehungen zwischen Organtätigkeit und Organdurchblutung. Naunyn-Schmiedeberg's Arch. exp. Path. Pharmak. **53**, 62—75 (1905).

(With H. H. MEYER) Über die Wirkung synthetischer, dem Adrenalin verwandter Stoffe. Naunyn-Schmiedeberg's Arch. exp. Path. Pharmak. **53**, 213—226 (1905).

(With T. ISHIZAKA) Über die Wirkung von Muskarin auf das nicht oder unzureichend gespeiste Froschherz und die Gegenwirkung von Kalziumsalz. Zbl. Physiol. **19**, 593—595 (1905).

1906

(With A. FRÖHLICH) Über vasokonstriktorische Fasern in der Chorda tympani. Zbl. Physiol. **20**, 229—232 (1906).

1907

Über Wirkungsweise und Indikation einiger diuretisch wirkender Mittel. Wien. klin. Wschr. 20, 1—5 (1907).

(With A. FRÖHLICH) Scheinbare Speisung der Nervenfaser mit mechanischer Erregbarkeit seitens ihrer Nervenzelle. (Nach Versuchen an Eledone moschata.) Zbl. Physiol. 21, 273—276 (1907).

1908

(With A. FRÖHLICH) Untersuchungen zur Physiologie und Pharmakologie des autonomen Nervensystems. I. Mitt. Über die Wirkung der Nitrite und des Atropins. Naunyn-Schmiedeberg's Arch. exp. Path. Pharmak. 59, 34—56 (1908).

(With E. NEUBAUER) Über Phlorhizindiurese und über die Beeinflussung der Phlorhizinzuckerausscheidung durch Diuretica. Naunyn-Schmiedeberg's Arch. exp. Path. Pharmak. 59, 57—63 (1908).

(With A. FRÖHLICH) Über vasoconstrictorische Fasern in der chorda tympani. Naunyn-Schmiedeberg's Arch. exp. Path. Pharmak. 59, 64—70 (1908).

(With D. JONESCU) Über eine spezifische Nierenwirkung der Digitaliskörper. Naunyn-Schmiedeberg's Arch. exp. Path. Pharmak. 59, 71—82 (1908).

Über eine neue Funktion des Pankreas und ihre Beziehung zum Diabetes mellitus. Naunyn-Schmiedeberg's Arch. exp. Path. Pharmak. 59, 83—94 (1908).

(With H. H. MEYER) Über Tetanusgift-Empfindlichkeit und Überempfindlichkeit. Naunyn-Schmiedebergs Arch. exp. Path. Pharmak. Suppl.-Bd. 355—365 (1908).

1910

(With A. FRÖHLICH) Über eine Steigerung der Adrenalinempfindlichkeit durch Cocain. Naunyn-Schmiedeberg's Arch. exp. Path. Pharmak. 62, 159—169 (1910).

(With G. MANSFELD) Über den Wirkungsmodus des Physostigmins. Naunyn-Schmiedeberg's Arch. exp. Path. Pharmak. 62, 180—185 (1910).

Pharmakologie und Klinik. Wien. klin. Wschr. 23, 273—278 (1910).

1911

Über Digitalistherapie. Mitt. Ver. Ärzte Steiermark 48, 379—387 (1911).

1912

Untersuchungen zur Physiologie und Pharmakologie des Herzvagus. I. Mitt.: Über den Einfluß von Chloralhydrat auf den Erfolg der Vagusreizung. — II. Mitt.: Über die Bedeutung des Calciums für die Vaguswirkung. — III. Mitteilung: Vaguserregbarkeit und Vagusgifte. Naunyn-Schmiedeberg's Arch. exp. Path. Pharmak. 70, 323—342, 343—350, 351—368 (1912).

1913

Über innere Sekretion. Mitt. Ver. Ärzte Steiermark 50, 71—83 (1913).

(With WESELKO) Über die Abhängigkeit experimentell-diabetischer Störungen von der Kationenmischung. Vorläufige Mitteilung. Münch. med. Wschr. 60, 690 (1913).

1914

(With O. WESELKO) Über den Einfluß der Thyreoidektomie auf die Wärmestichreaktion bei Kaninchen. Zbl. Physiol. 28, 197 (1914).

Über die Folgen der Nebennierenexstirpation beim Frosch. (Dtsch. Physiol. Ges., 2. bis 5. Juni 1914.) Abstr.: Zbl. Physiol. 28, 727 (1914).

(With W. GETTWERT) Über die Folgen der Nebennierenexstirpation. I. Mitt. Untersuchungen am Kaltblüter. Pflügers Arch. ges. Physiol. 158, 29—40 (1914).

(With O. WESELKO) Über den Kohlenhydratumsatz des isolierten Herzens normaler und diabetischer Tiere. Pflügers Arch. ges. Physiol. 158, 155—188 (1914).

1917

Über die Calciumwirkung und Calciumtherapie. Mitt. Ver. Ärzte Steiermark **54**, 53—59, 63—71 (1917).

Über die Entstehung von functionellen Anpassungen im individuellen Dasein. Naturwissenschaften **5**, 501—507 (1917).

Über den Zusammenhang von Digitalis- und Kalziumwirkung. Münch. med. Wschr. **64**, 1003—1004 (1917).

Über den Zusammenhang zwischen Digitalis- und Kalziumwirkung. Naunyn-Schmiedeberg's Arch. exp. Path. Pharmak. **82**, 131—158 (1917/18).

1918

Über den Zusammenhang zwischen Digitalis- und Kalziumwirkung. III. Mitt. Naunyn-Schmiedeberg's Arch. exp. Path. Pharmak. **83**, 366—380 (1918).

Über Spontanerholung des Froschherzens bei unzureichender Kationenspeisung. II. Mitt. Ein Beitrag zur Wirkung der Alkalien aufs Herz. Pflügers Arch. ges. Physiol. **170**, 677—695 (1918).

Zur Frage der Verwertbarkeit der Glukose bei Diabetes. Ther. Mh. **32**, 350—354 (1918).

1919

(With H. LIEB) Über Spontanerholung des Froschherzens bei unzureichender Kationenspeisung. III. Mitt. Quantitative mikroanalytische Untersuchungen über die Ursache der Calciumabgabe von seiten des Herzens. Pflügers Arch. ges. Physiol. **173**, 152—157 (1919).

Über Wertung und Wirkung von Werken der bildenden Kunst. Z. ästhet. Kunstwiss. **14**, 239—252 (1919).

1921

Über die Beziehungen zwischen Herzmittel- und physiologischer Kationenwirkung. IV. Mitt. Über Nichtelektrolytwirkung aufs Herz. — V. Mitt. Über die Wirkung von Lipoiden auf die Hypodynamie und deren Beziehung zum Kalium. Pflügers Arch. ges. Physiol. **187**, 105—122, 123—131 (1921).

Über die Beziehungen zwischen Herzmittel- und physiologischer Kationenwirkung. VI. Mitt. Über die Kaliumcontractur. Pflügers Arch. ges. Physiol. **188**, 87—97 (1921).

Über humorale Übertragbarkeit der Herznervenwirkung. I. Mitt. Pflügers Arch. ges. Physiol. **189**, 239—242 (1921).

1922

Über humorale Übertragbarkeit der Herznervenwirkung. II. Mitt. Pflügers Arch. ges. Physiol. **193**, 201—213 (1922).

Weitere Untersuchungen über humorale Übertragbarkeit der Herznervenwirkungen. Klin. Wschr. **1**, 22—23 (1922).

Über humorale Übertragbarkeit der Herznervenwirkung. Naturwissenschaften **10**, 52—55 (1922). Also: Verh. dtsch. pharmak. Ges. 1921, Nr 1, S. XXIV. In: Naunyn-Schmiedeberg's Arch. exp. Path. Pharmak. **92** (1922).

Sulla trasmissibilità umorale dell'azione dei nervi cardiaci. Biochim. Terpa. sper. **9**, 175—191 (1922).

(With E. GEIGER) Über Änderung des Cholingehaltes der Froschmuskulatur durch elektrische Reizung. Biochem. Z. **127**, 174—180 (1922).

(With E. GEIGER) Versuche über die Glucosepermeabilität der Leber. Klin. Wschr. **1**, 1210 (1922).

(With J. SOLTI) Über die Wirkung von Pilocarpin und Atropin auf den isolierten Krötenmuskel und ihre Abhängigkeit von der Ionen-Mischung. Klin. Wschr. **1**, 2046 (1922).

1923

Die Ursache der Unempfindlichkeit des Krötenherzens für die Contracturwirkung der Digitalisstoffe. Pflügers Arch. ges. Physiol. **198**, 359—366 (1923).

(With E. GEIGER) Versuche über die Glucosepermeabilität der Leber. Pflügers Arch. ges. Physiol. **198**, 633—643 (1923).

Über Steuerungen von Funktionen im Tierkörper. Naturwissenschaften **11**, 117—123 (1923).

(With J. SOLTI) Über die Wirkung von Pilocarpin und Atropin auf den quergestreiften Muskel. Naunyn-Schmiedeberg's Arch. exp. Path. Pharmak. **97**, 272—284 (1923).

Über die humorale Übertragbarkeit der Herznervenreizungswirkung. Bemerkungen zu dem Aufsatz von H. J. HAMBURGER in Jg. 2, Nr. 28 dieser Wochenschrift. Klin. Wschr. **2**, 1840—1841 (1923).

1924

(With G. SINGER) Über die Wirkung des Jods auf die Atmung isolierter Zellen. Wien. med. Wschr. **74**, 328—332 (1924).

Weiteres über humorale Übertragbarkeit der Herznervenwirkung. Klin. Wschr. **3**, 680—681 (1924).

Über humorale Übertragbarkeit der negativ chronotropen und negativ dromotropen Vaguswirkung. Klin. Wschr. **3**, 1078 (1924).

Über humorale Übertragbarkeit der Herznervenwirkung. III. Mitt. Pflügers Arch. ges. Physiol. **203**, 408—412 (1924).

Über humorale Übertragbarkeit der Herznervenwirkung. IV. Mitt. Pflügers Arch. ges. Physiol. **204**, 361—367 (1924).

Über humorale Übertragbarkeit der Herznervenwirkung. V. Mitt. Die Übertragbarkeit der negativ chrono- und dromotropen Vaguswirkung. Pflügers Arch. ges. Physiol. **204**, 629—640 (1924).

(With E. NAVRATIL) Über humorale Übertragbarkeit der Herznervenwirkung. VI. Mitt. Der Angriffspunkt des Atropins. Pflügers Arch. ges. Physiol. **206**, 123—134 (1924).

(With E. NAVRATIL) Über humorale Übertragbarkeit der Herznervenwirkung. VII. Mitt. Pflügers Arch. ges. Physiol. **206**, 135—140 (1924).

1925

(With H. HÄUSLER) Insulin und die Zuckerverteilung zwischen flüssigen und nicht flüssigen Systemen. (Kurze Mitteilung.) Biochem. Z. **156**, 295—299 (1925).

(With H. HÄUSLER) Zur Frage der Wirkungsweise des Insulins. I. Mitt. Insulin und die Glucoseverteilung zwischen flüssigen und nichtflüssigen Systemen. Pflügers Arch. ges. Physiol. **210**, 238—279 (1925).

1926

(With S. DIETRICH u. H. HÄUSLER) Weiteres über Insulinwirkung und Diabetes. Klin. Wschr. **5**, 414 (1926).

(With E. NAVRATIL) Über das Schicksal des Vagusstoffs und des Acetylcholins im Herzen Klin. Wschr. **5**, 894 (1926).

(With E. NAVRATIL) Über den Mechanismus der Vaguswirkung von Physostigmin und Ergotamin. Klin. Wschr. **5**, 894—895 (1926).

Kritische Bemerkungen zu L. ASHERS Mitteilungen. Pflügers Arch. ges. Physiol. **212**, 695—706 (1926).

(With H. HÄUSLER) Untersuchungen über Diabetes und Insulinwirkung. V. Mitt. Pflügers Arch. ges. Physiol. **214**, 370—379 (1926).

(With E. NAVRATIL) Über humorale Übertragbarkeit der Herznervenwirkung. X. Mitt. Über das Schicksal des Vagusstoffs. Pflügers Arch. ges. Physiol. **214**, 678—688 (1926). — XI. Mitt. Über den Mechanismus der Vaguswirkung von Physostigmin und Ergotamin. Pflügers Arch. ges. Physiol. **214**, 689—696 (1926).

Über die Wirkung des Insulins und des Insulin-Antagonisten des diabetischen Blutes. Nach Versuchen mit S. DIETRICH und H. HÄUSLER. Wien. klin. Wschr. 39, 1074—1076 (1926).
(With S. DIETRICH) Untersuchungen über Diabetes und Insulinwirkung. VII. Mitt. Pflügers Arch. ges. Physiol. 215, 78—94 (1926).
I. Über das Schicksal des Vagusstoffes im Herzen. II. Über den Mechanismus der Vaguswirkung von Eserin und Ergotamin. Abstracts of Communications to the XIIth Internat. Physiological Congr. held at Stockholm, August 3—6, 1926. Skand. Arch. Physiol. 49, 172 (1926).
Über den Vagusstoff. Naturwissenschaften 14, 994—995 (1926).

1927

Über Strukturfixierung der Glucose und ihre Bedeutung für das Glucoseschicksal. Naturwissenschaften 15, 93—94 (1927).
(With M. L. PATRIZI) Intorno al sopposto "ormone vagale" di LOEWI. Nuovi esperimenti sui cuori coniugati dei cani di grossa taglia e nuovi argomenti.) Bull. Sci. med. 5, 25—43 (1927).
(With H. HÄUSLER) Über die alimentäre Hyperglykämie. Klin. Wschr. 6, 313 (1927).
(With S. DIETRICH) Über den Phlorhizindiabetes. Klin. Wschr. 6, 313 (1927).
(With S. DIETRICH and H. HÄUSLER) Weiteres über den insulinantagonistischen Stoff. Klin. Wschr. 6, 856 (1927).
(With H. HÄUSLER) Weiteres über alimentäre Hyperglykämie. Klin. Wschr. 6, 856 (1927).
(With H. HÄUSLER) Untersuchungen über Diabetes und Insulinwirkung. IX. Mitt. Über das Auftreten des insulinantagonistischen Stoffes im Blut nach Pankreasexstirpation. X. Mitt. Weitere Wirkungen des insulinantagonistischen Hormones „Glykämin" und ihre Bedeutung für den Mechanismus des Diabetes. Naunyn-Schmiedeberg's Arch. exp. Path. Pharmak. 123, 56—62, 63—71 (1927).
(With H. HÄUSLER) Über hormonale Vorgänge nach Glucosezufuhr. I. Mitt. Über Insulinsekretion nach subcutaner Glucosezufuhr, ihre quantitative Verfolgung und ihren Mechanismus. — II. Mitt. Über Insulin- und Glykäminsekretion nach peroraler Glucosezufuhr. — III. Mitt. Versuche am Hungertier. Naunyn-Schmiedeberg's Arch. exp. Path. Pharmak. 123, 72—87, 88—119, 120—128 (1927).
Glykämin und Insulin. I. Klin. Wschr. 6, 2169—2176 (1927). — ZUELZER, G.: Bemerkungen. Klin. Wschr. 7, 312 (1928). — LOEWI, O.: Erwiderung. Klin. Wschr. 7, 312 (1928).

1928

Zur Frage nach dem Mechanismus der Glykogenolysesteigerung beim Diabetes. Klin. Wschr. 7, 311—312 (1928).
(With S. DIETRICH) Insulin und Glykämin. II. Klin. Wschr. 7, 629—634 (1928).

1929

Insulin und Glykämin. III. Klin. Wschr. 8, 391—393 (1929).
(With H. HÄUSLER) Über die Glucosefixation durch Blutkörperchen. Bemerkungen zur gleichnamigen Arbeit von HÄGLER, THOMANN und ÜBERRACK. Biochem. Z. 214, 229—230 (1929).
Bemerkungen zur Receptorenfunktion im Verdauungskanal. Dtsch. med. Wschr. 55, 1751—1752 (1929).

1930

Hermann Pfeiffer. Klin. Wschr. 9, 575 (1930).
(With E. ENGELHART) Fermentative Acetylcholinspaltung im Blut und ihre Hemmung durch Physostigmin. Naunyn-Schmiedeberg's Arch. exp. Path. Pharmak. 150, 1—13 (1930).

(With E. Engelhart) Über die Vagusstoffzerstörung im Herzen. Arch. int. Pharmacodyn. **38**, 287—291 (1930).

1931

Über „neuro-humorale" Auslösungen im Organismus. Jkurse ärztl. Fortbild. **22**, H. 1,
   1—11 (1931). Also: Karlsbad. ärztl. Vortr. **12**, 325—342 (1931).
Der fehlende Zusammenhang zwischen Herznervenreizung und Bildung eines humoralen
   Vagusstoffes. Bemerkungen zur Mitteilung von L. Asher und N. Scheinfinkel.
   Klin. Wschr. **10**, 1312 (1931).

1932

Eröffnungsansprache. Verh. dtsch. Ges. inn. Med. **44**, 13—16 (1932).
Eröffnungsansprache. Verh. Dtsch. Pharmak. Ges. 1932, pp. 17—20.

1933

(With E. Pichler) Über den Glykogenstoffwechsel des Muskels und seine nervöse Beeinflussung. III. Mitt. Ein propriozeptiver glykogenolytischer Reflex. Pflügers Arch.
   ges. Physiol. **233**, 51—56 (1933).

1934

(With M. Fluch and H. Greiner) Hypophysenvorderlappen und Glykogenolyse. Klin.
   Wschr. **13**, 883—884 (1934). Also: Naunyn-Schmiedeberg's Arch. exp. Path. Pharmak. **177**, 167—176 (1935).

1935

The Ferrier Lecture on problems connected with the principle of humoral transmission
   of nervous impulses. Proc. roy. Soc. B **118**, 299—316 (1935).

1936

Quantitative und qualitative Untersuchungen über den Sympathicusstoff. Zugleich
   XIV. Mitt. über humorale Übertragbarkeit der Herznervenwirkung. Pflügers Arch.
   ges. Physiol. **237**, 504—514 (1936).

1937

Über den Adrenalingehalt des Säugerherzens. Arch. int. Pharmacodyn. **57**, 139—140
   (1937).
Die chemische Übertragung der Nervwirkung. Schweiz. med. Wschr. **67**, 850—855
   (1937).
(With R. Hagen, H. Kohn and G. Singer) Über eine wasserunlösliche Zustandsform des
   Acetylcholins im Zentralnervensystem des Frosches. Pflügers Arch. ges. Physiol. **239**,
   430—439 (1937).

1938

Über den Mechanismus der Stoffwechselwirkung des Hypophysenvorderlappens. (Pavlov
   Memorial.) Fiziol. Zh. SSSR. **24**, 241—244 (1938). Abstr.: Kongr.-Zbl. ges. inn. Med.
   97, 239 (1938).
(With H. Hellauer) Vergleich der Acetylcholinausbeute aus dem Zentralnervensystem
   des Frosches bei Extraktion mit HCl-Alkohol und Trichloressigsäure sowie mittels
   Kochen. Pflügers Arch. ges. Physiol. **240**, 449—457 (1938).
(With H. Hellauer) Über das Acetylcholin in peripheren Nerven. Pflügers Arch. ges.
   Physiol. **240**, 769—775 (1938).

1939

(With W. Dulière) Liberation of potassium by acetylcholine in the central nervous
   system. Nature (Lond.) **144**, 244 (1939).

1944

(With G. L. CANTONI) Inhibition of cholinesterase activity of nervous tissues by eserine
   in vivo. J. Pharmacol. exp. Ther. **81**, 67—71 (1944).

1945

(With S. BLUMENFELD) Digitalis and calcium. J. Pharmacol. exp. Ther. **83**, 96—99 (1945).
The Edward Gamaliel Janeway Lectures: Aspects of the transmission of nervous impulse.
   Mediation in the peripheral and central nervous system. J. Mt Sinai Hosp. **12**,
   803—816 (1945). — Theoretical and clinical implications. J. Mt Sinai Hosp. **12**,
   851—865 (1945).
Chemical transmission of nerve impulses. Sci. Progr., Ser. IV (1945).

1946

The antagonism of tetraethylammonium against heart depressants. J. Pharmacol.
   exp. Ther. **88**, 136—141 (1946).

1948

Control of internal secretions. Proc. R. Virchow med. Soc. **7**, 1—14 (1948).

1949

On the hypersensitivity of denervated structures. Confin. neurol. (Basel) **9**, 58—63
   (1949).
On the antagonism between pressor and depressor agents in the frog's heart. J. Pharmacol. exp. Ther. **96**, 295—304 (1949).

1950

Smasnak om farmakologi. Med. Forum (Kbh.) **3**, 225—234 (1950).
Problems in the field of adrenal function. Proc. R. Virchow med. Soc. **9**, 101—110
   (1950).

1952

Ernst Peter Pick, the scientist. J. Mt Sinai Hosp. **19**, 12—13 (1952).
On the action of fluoride on the heart of Rana pipiens; preliminary note. J. Mt Sinai
   Hosp. **19**, 1—3 (1952).

1953

(With E. P. PICK and J. WARKANY) Alfred Fröhlich, 1871—1953. Science **118**, 314 (1953).

1954

Introduction to Symposium on Neurohumoral Transmission, Physiological Society of
   Philadelphia, Sept. 11 and 12, 1953. Pharmacol. Rev. **6**, 3—6 (1954).
Reflections on the study of physiology. Ann. Rev. Physiol. **16**, 1—10 (1954).

1955

On the mechanism of the positive inotropic action of fluoride, oleate and calcium on the
   frog's heart. J. Pharmacol. exp. Ther. **114**, 90—99 (1955).
Salute to Henry Hallett Dale. Brit. med. J. **1955**, I. 1356—1357.

1956

On the intraneural state of acetylcholine. Experientia (Basel) **12**, 331—333 (1956).

1957

On the background of the discovery of neurochemical transmission. J. Mt Sinai Hosp.
   24, 1014—1016 (1957).

1958

A scientist's tribute to art. Essays in honour of Hans Tietze. Gaz. Beaux Arts, 1958.

The Cerebrospinal Fluid

By

HUGH DAVSON*

With 8 Figures

Table of Contents

* Medical Research Council, Department of Physiology, University College London.

I. Introduction

In recent years the more physiological aspects of the cerebrospinal fluid have aroused interest, with the result that some progress in elucidating the mechanism of its formation and circulation, and its relationships with the central nervous parenchyma, has been made. The position is still very obscure, however, and it is the purpose of this review to collect the facts from a very scattered literature and to emphasize the gaps, in the hope that they will invite further research. In general, I have used as my starting point in time the years 1955 to 1956, the period when an earlier and more comprehensive review (DAVSON 1956) was going through the press; this has not precluded the quotation of earlier work, however, when this was relevant. In writing this, I have of course been able to benefit by the series of articles in the Ciba Symposium held in London in 1957 and the scholarly reviews of the blood-brain barrier by BAKAY (1956) and DOBBING (1961). I have tried to be comprehensive, but the wide area of the literature over which the published work is scattered means, unquestionably, that I have overlooked many significant contributions[1].

II. The blood-cerebrospinal fluid and blood-brain barriers

The concept developed by WEED (1914), namely that the cerebrospinal fluid is secreted by the choroid plexuses in the lateral, IIIrd and IVth ventricles, and flows into the cisterna magna to be finally drained away in the arachnoid villi of the venous sinuses, has not been seriously challenged. On this basis, the passage of material from blood to cerebrospinal fluid represents the passage of material from the blood in the choroid plexuses into the primary secretion of fluid poured into the ventricles; secondly it includes materials from the capillaries of the nervous tissue through the extracellular spaces of this tissue, and into the cerebrospinal fluid as it passes through the ventricles and subarachnoid spaces. For simplicity the system may be illustrated by Fig. 1. It is obvious from this that the blood-cerebrospinal fluid and blood-brain barriers are closely related, and the extent and nature of this relationship have come in for some experimental work and some discussion.

A. Blood-cerebrospinal fluid barrier

It was shown earlier (DAVSON 1955, 1956) that the uptake of material from blood by the cerebrospinal fluid (csf) in many cases followed a simple equation of the form:

$$\frac{d\,C_{csf}}{d\,t} = k_{In}\,C_{Plasma} - k_{Out}\,C_{csf}$$

---

[1] In addition I am indebted to D. P. RALL and C. G. ZUBROD for showing me the manuscript of a review discussing the passage of drugs into and out of the central nervous system. The review, entitled "Mechanisms of Drug Absorption and Excretion", has since been published in Annual Reviews of Pharmacology 2, 109—128, 1962.

the values of k_{In} and k_{Out} increasing with increasing lipid-solubility. Studies from Brodie's laboratory on a variety of drugs have confirmed and emphasized

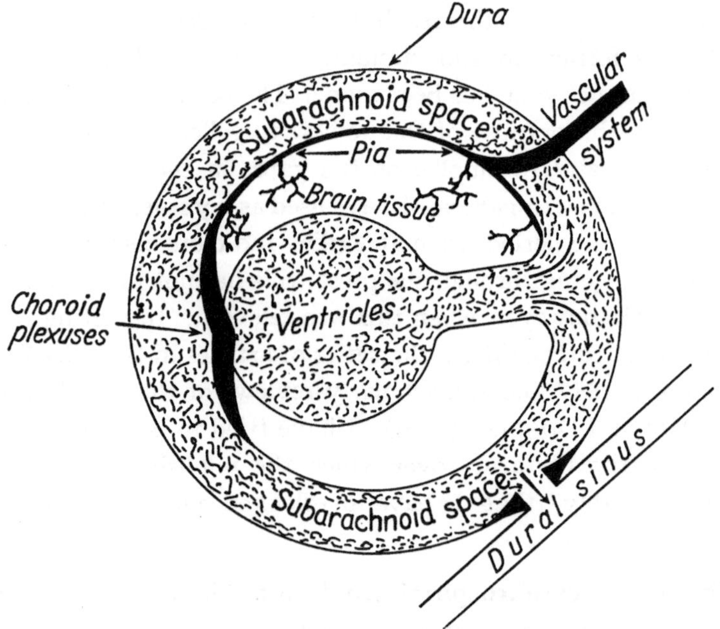

Fig. 1. Schematic illustration of relationships of cerebrospinal fluid, blood and nervous tissue

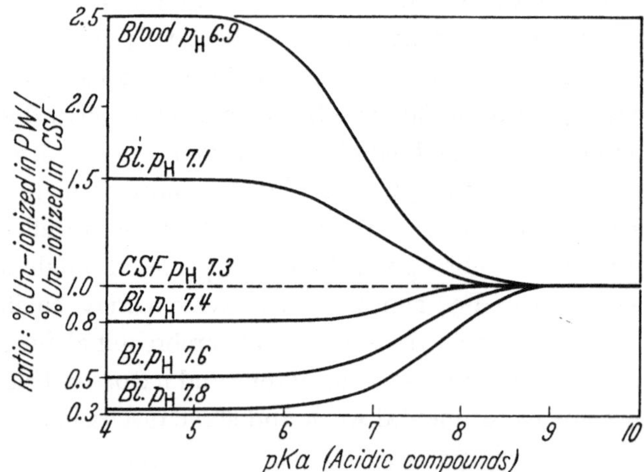

Fig. 2. The theoretical plasma water/csf ratio of unionized compound as a function of varying p_H gradient. R_{csf} is directly proportional to the ratio: per cent unionized molecular species in plasma water/ per cent of unionized molecular species in csf. Hence this graph demonstrates the calculable effect of p_H gradient between blood and csf on the R_{csf} of anionically dissociating compounds of varying pKa's. Csf p_H is assumed to be constant at 7.3 while blood p_H varies from 6.9 to 7.8. For distribution of cationically dissociating compounds the abcissa would be 14 minus pKa of the compound. The ratio of per cent unionized compound at the points on the ordinate corresponding to the various blood p_H's would be the reciprocal of those values shown. (From RALL, STABENAU and ZUBROD 1959)

the importance of lipid-solubility and, where the drugs are weak acids or bases, of the degree of dissociation (BRODIE and HOGBEN 1957; BRODIE, KURZ and SCHANKER 1960) whilst RALL, MOORE, TAYLOR and ZUBROD (1961) have extended these results from dog to man. With the relatively lipid-soluble and undissociated substances like the substituted thioureas, the final steady-state distribution between plasma and cerebrospinal fluid achieved was unity, i.e. k_{In} could be equated to k_{Out}. With the weak bases, however, it was shown by RALL, STABENAU and ZUBROD (1959), STABENAU, WARREN and RALL (1959) and ZUBROD and RALL (1959) that the steady-state distribution was a function of the pH of plasma and cerebrospinal fluid, as would be expected on theoretical grounds, the distribution ratio R_{csf} being given by the equation:

$$R_{csf} = \frac{1 + 10^{\text{PH csf} - pK_a}}{1 + 10^{\text{PH pl} - pK_a}}.$$

Thus, as Fig. 2 illustrates, the discrepancy of R_{csf} from unity may be quite large when the pK_a is in the region of 6 or less.

B. Blood-brain barrier

To continue with these kinetically simple substances, it was shown by DAVSON (1955) that their penetration into the nervous tissue was also determined by lipid-solubility and that, in general, the water of this tissue came into equilibrium with the plasma rather more rapidly than did the cerebrospinal fluid, so that the cerebrospinal fluid tends to act as a "sink" for material entering the nervous tissue. This is illustrated by Fig. 3 from a recent paper by DAVSON, KLEEMAN and LEVIN (1962) where it is seen that the concentration of ethyl thiourea, a fairly lipid-soluble substance, is at all times higher in the nervous tissue than in the cerebrospinal fluid when a constant concentration of this sub-

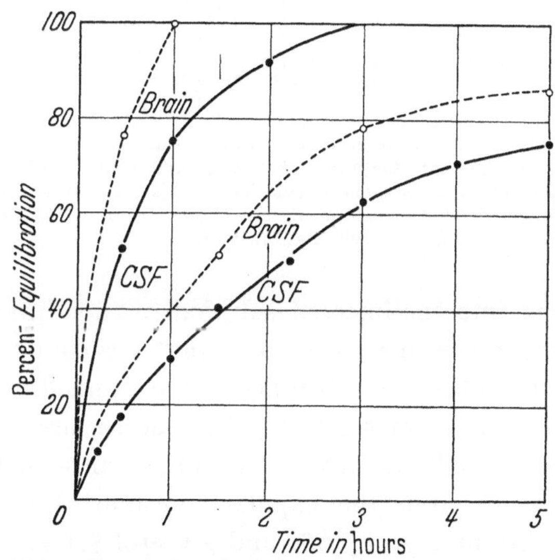

Fig. 3. Penetration of ethyl thiourea (upper curves) and thiourea (lower curves) into cerebrospinal fluid and brain

stance is maintained in the plasma. This confirms the supposition, made in the previous section, that the uptake of material by the cerebrospinal fluid from the blood occurred both directly from the choroid plexuses and indirectly from the adjacent nervous tissue. Whether this is the situation with all penetrating substances is, of course, not proved; as we shall see, this point is of importance in assessing the extracellular space of nervous tissue (p. 28). The importance of lipid-solubility in determining penetration of the blood-brain barrier has been sustained by work from Brodie's laboratory (BRODIE and HOGBEN 1957; MAYER, MAICKEL and BRODIE 1959; BRODIE et al. 1960). Fig. 4 from MAYER et al. demonstrates the applicability of the simple kinetic equation for penetration mentioned earlier, and shows the importance of lipid-solubility since the heptane/water partition coefficients are 0.001, 0.004, 0.005 and 0.01 for salicylic acid, N-acetyl-4-aminoantipyrine, barbital and acetanilide respectively. These authors also confirmed that the penetration of these drugs into nervous tissue was always more rapid than that into the cerebrospinal fluid.

Finally CRONE (1961) has described the results of an elaborate series of experiments on the penetration of organic substances into the brain of the dog. The technique was new, so far as the brain was concerned, and consisted in analysing the venous effluent from the brain (superior sagittal sinus) during the course of a steady intracarotid infusion; simple dilution effects, due to admixture of blood not containing the injected material, were allowed for by incorporat-

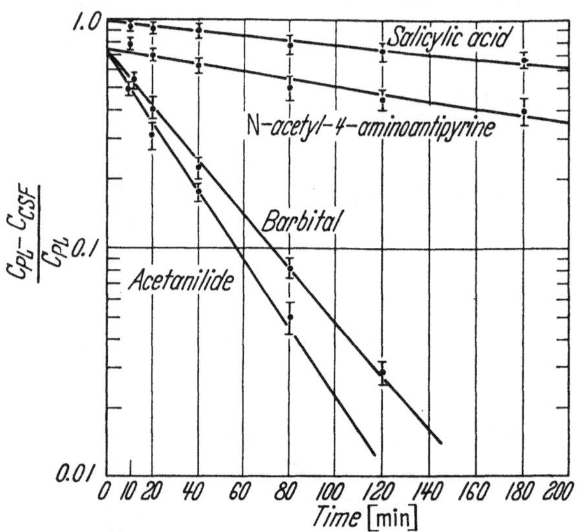

Fig. 4. Kinetics of entry of various drugs into csf. Each point represents the mean of 4 to 6 experiments. Vertical bars represent standard deviation. C_{pl} is concentration of unbound drug in plasma water. (From MAYER, MAICKEL and BRODIE 1959)

ing into the injected solution a "non-diffusible reference substance", i.e. a substance like Evans blue that was confined to the cerebral capillary system. The following substances were studied: ethanol, propanol, butanol, inulin, sucrose, fructose, antipyrin, glucose, urea, thiourea, glycerol, ethylene glycol. The results confirmed the importance of lipid-solubility in determining the escape from the capillary system; thus propanol gave 90% extraction, ethylene glycol 9.6% and glycerol 5.4%. Glucose, as one might expect, was exceptional in showing a high degree of permeability in spite of its low lipid-solubility, and there seems no doubt that facilitated diffusion is operative here.

C. Extracellular space of brain

We have so far discussed the behaviour of some relatively lipid-soluble substances; for these the whole of the tissue-water is presumably available because the lipid-solubility confers on the molecules the power not only to leave the vascular system at a high rate but also to penetrate all the cells. When a number of less lipid-soluble substances, such as sucrose, ferrocyanide, iodide, thiocyanate, p-aminohippurate, are studied, the rates of appearance in both cerebrospinal fluid and nervous tissue are remarkably small and, of course, it was the early qualitative studies on such substances that led to the concepts of blood-cerebrospinal fluid and blood-brain barriers. To indicate the barriers quantitatively, we may give the values of the ratio: Concentration in csf/concentration in plasma reached when a constant level of the substances has been maintained in the plasma for two hours. The values are as follows:

Ethyl alcohol	Ethyl-thiourea	^{24}Na	Sucrose	^{131}I	CNS
1.0	0.925	0.40	0.010	0.012	0.095

So far as the brain is concerned the results may be indicated by the volume of distribution of the given substance at two intervals after establishing a constant concentration in the plasma; these may be compared with skeletal muscle as in Fig. 5. It is seen that, whereas the volumes of distribution in the muscle become equal to the "chloride-space" in 6 min (or possibly less time), even after two hours these volumes are only some 4 % of the tissue and are far removed from the chloride-space.

Since these substances are typically extracellular, their very small volumes of distribution after two hours may be due to the very high barrier between the blood and the extracellular space, corresponding to a similar barrier between blood and cerebrospinal fluid, or they may be due to a very small volume of the extracellular space in nervous tissue.

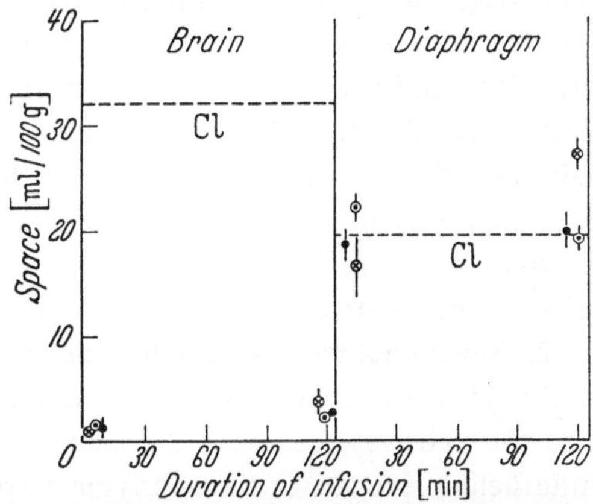

Fig. 5. Penetration of PAH, ⊗, sucrose, ●, and ^{131}I, ○, into brain and diaphragm muscle of rabbits infused with the test substance for 6 or 120 min. The ordinate refers to the volume of tissue water (per 100 g tissue) that contains the same concentration of the test substance as in a plasma dialysate. Each point represents the average of determinations in 4—10 animals; s. e. shown as vertical lines. (From DAVSON and SPAZIANI 1959)

Thus, on the basis of the second explanation, the failure of extracellular substances to enter the nervous tissue at appreciable rates, and in appreciable amounts, is due not to the presence of a barrier to diffusion from the blood, but to the absence of an appreciable volume of extracellular fluid with which to come into equilibrium (cf. EDSTRÖM 1958). Thus, the magnitude of the extracellular space of brain and spinal cord is a topic of extreme interest in the interpretation of "barrier phenomena".

1. **Anatomical studies.** Electron-microscopical studies of the central nervous system have been published by WYCKOFF and YOUNG (1956), LUSE (1956), SCHULTZ, MAYNARD and PEASE (1957), FERNANDEZ-MORÁN (1957), FARQUHAR

and Hartmann (1957) and Horstmann and Meves (1959) and others. There is general agreement that the tissue is made up of an aggregate of closely packed cellular elements — glial cells and neurones — the space between any two cells being very regular and, as seen in the dried and fixed specimens, at the most some 200 Å thick, i.e. no greater than that between the "tightly packed cells of an epithelium". Of these studies the one that has been discussed most exhaustively from the viewpoint of the extracellular space and blood-brain barrier is that of Horstmann and Meves. As they point out, the value of the computed extracellular space, expressed as a percentage of the whole tissue, will increase the smaller the cellular elements, so that on the basis of a 200 Å space between these elements the space will be as high as 30 % if the diameters of the cellular elements average 0.1μ, but will be as low as 5—7 % with diameters of 0.5—0.75 μ and with large aggregations of perikarya the space will be even smaller. Direct measurements from photos of sections gave values of 1—15 % depending on the area examined, with an average for the whole tissue of about 5 %. The electron-microscopical evidence thus provides a plausible basis for dismissing the blood-brain barrier phenomenon as an indication only of the limited extracellular space of the tissue, rather than the existence of a definite diffusion-resistance offered by the membranes separating blood from this extracellular space. Before discussing further the ultrastructure of the nervous tissue, therefore, we may consider the justification, or otherwise, of this rejection of a barrier between blood and extracellular space.

2. Functional studies. Implicit in any definition of the blood-brain barrier is the presence of a resistance to diffusion from the vascular compartment into the extravascular one; the locus of this resistance may be the capillary endothelium proper, or the overlying astrocytic and oligodendroglial processes that envelop the endothelial cells; it does not matter which. If there is a resistance interposed between vascular and extravascular compartments, it may be circumvented by applying a solution of the substances in which we are interested, e.g. iodide, thiocyanate, or sucrose, to the tissue directly. This may be done *in vitro* by soaking pieces of tissue in a Krebs-Ringer medium containing the substances and measuring the uptake at different times (Davson and Spaziani 1959). Some results of this type of experiment are shown in Fig. 6, and leave us in no doubt that such "extracellular" substances as sucrose, ^{131}I, and p-aminohippurate (PAH) can pass into the nervous tissue rapidly (as rapidly as into muscle such as diaphragm if allowance is made for the different thickness), to give volumes of distribution that are quite large, e.g. 10—15 % for sucrose and considerably higher for [131]I. Since these substances are unable to reach this volume of distribution when presented by way of the blood-stream, the results leave us in little doubt as to the presence of a barrier to diffusion from vascular to extravascular compartments. It can be argued, of course, that the tissues studied *in vitro* were abnormal, so that

an *in vivo* study would be preferable; this has been recently carried out (DAV-
SON, KLEEMAN and LEVIN 1961) by barbotage of the subarachnoid space with
a Ringer's solution containing sucrose and PAH. Repeated replacement of
the cerebrospinal fluid of the rabbit with the solution ensured that the subarachnoid space of the cord (not the brain, however) was filled with this, so
that the underlying cord was exposed to an approximately constant concentration of the two extracellular "tags". After two hours the cord was removed
and analysed to give an average figure of 12 % for the sucrose-space and 10 %

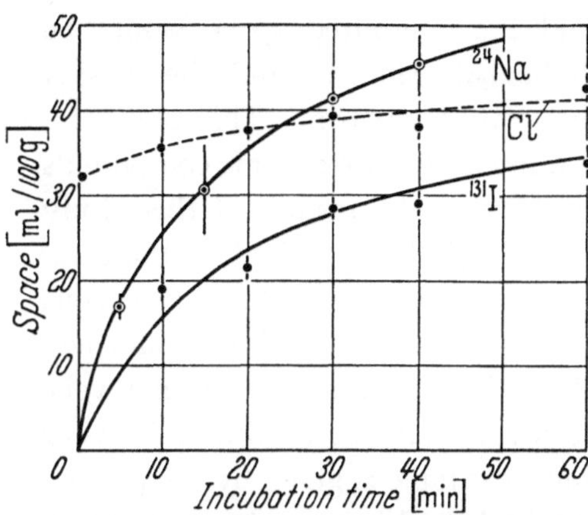

Fig. 6a. Penetration of ²⁴Na and ¹³¹I, and changes
in chloride space, in pieces of cerebral hemisphere
incubated at 40° C. Ordinate as in Fig. 5. Each
point represents the average of 6—10 incubations;
s.e. shown as vertical lines. (From DAVSON and
SPAZIANI 1959)

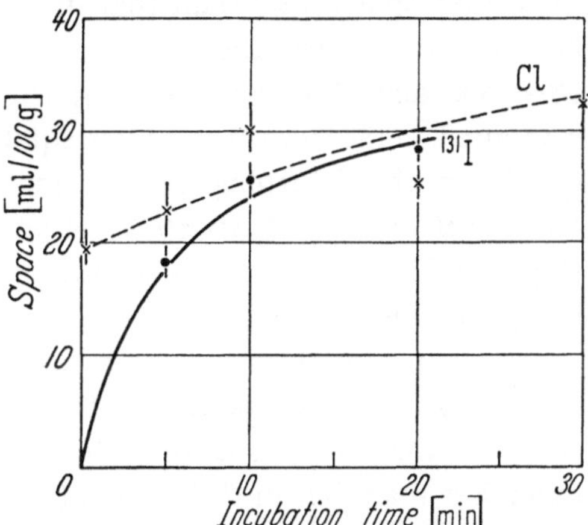

Fig. 6b. Penetration of ¹³¹I, and changes in chloride
space, in diaphragms incubated at 40° C; ordinate as
in Fig. 5. Each point represents the average of 4—7
incubations; s.e. shown as vertical lines. (From
DAVSON and SPAZIANI 1959)

for the PAH-space. Thus the experiments demonstrate unequivocally that
there is a space available for the penetration of sucrose amounting to at least
12 % of the tissue, yet this is not occupied by sucrose when presented by way
of blood even after 2 hours. The concept of the blood-brain barrier is thus
sustained, but of course the anomaly between the anatomical and functional
studies remains.

3. Quantitative estimates of the extracellular space. Clearly, the most
physiological method of measuring the extracellular space would be to maintain
a known concentration of an extracellular tag in the blood for a very long time
so that equilibrium between plasma and the space could take place; analysis
of blood and tissue would then give the magnitude of the space. Because of
the existence of the blood-brain barrier, however, the time would have to be
very long indeed for equilibrium to be reached; moreover, because of a variety
of complicating factors, it may well be that achievement of a true equilibration
will be impossible. Thus, with $^{35}SO_4$ there is the danger that some of the
sulphur will be incorporated into organic sulphur-compounds; with inulin that
it will be taken up by the phagocytic elements, and so on. BARLOW, DOMEK,

Goldberg and Roth (1961) injected inorganic $^{35}SO_4$ into ureter-ligated animals and after 6 hr analysed brain and blood; they computed a space of only 2—4% but it is questionable whether they were measuring a true equilibrium space because of the "sink-action" of the cerebrospinal fluid. Thus the concentration in this fluid was only 9% of that in the plasma so that, because it is well established that sulphate and other extracellular tags can exchange between nervous tissue and cerebrospinal fluid, it seems very likely that the extracellular fluid would acquire a concentration that was closer to that in the cerebrospinal fluid than that in the plasma; if the extracellular fluid were in equilibrium with the cerebrospinal fluid the computed space would be 18—36% and clearly the true space will be between this and the 2—4% computed on the basis of complete equilibrium with the plasma. This, of course, is to ignore the disturbing effects of incorporation of ^{35}S into organic compounds (Boström and Odeblad 1953; Ringerty 1956). The results of this study thus serve to emphasize the disturbing effect of the "sink-action" of the cerebrospinal fluid when the penetration of extracellular tags into the nervous tissue is studied. This fluid is a circulating one, being drained and reformed continuously; the extracellular tags penetrate into it from the plasma very slowly so that at the steady state the concentration in the fluid may be very much less than that in the plasma (an extreme example is given by the plasma proteins, their concentration in the cerebrospinal fluid being only about 1/300 that in the plasma). Consequently we cannot expect the concentration of an extracellular tag, such as inulin or sulphate, to rise to that in the plasma, however long is allowed for the equilibration to occur. Similar criticisms must apply to the studies of Woodbury, Timiras, Koch and Ballard (1956) on the inulin- and $^{35}SO_4$-spaces in the brains of rats, of Reed and Woodbury (1960) on the sucrose-space, and of Morrison (1959) on the inulin-space of dog and rat brain; this last author gave a single injection of inulin and analysed the tissue and cerebrospinal fluid 3—7 hours later; his computed space was 5%, but the concentration in the cerebrospinal fluid was only one twentieth that in the plasma, so that there must have been a considerable "sink-action".

Streicher (1961) made the interesting observation that the thiocyanate-space increased with increasing concentration of this ion in the blood of the rat; this result, at first sight surprising, is in accordance with the view adopted above of the cerebrospinal fluid acting as a sink for the nervous tissue. Thus Streicher, Rall and Gaskins (1961) found that the rate of penetration of thiocyanate into the cerebrospinal fluid, and the magnitude of the steady-state distribution ratio, increased with increasing plasma concentration. This is due to the saturation of an absorptive mechanism that actively removes thiocyanate from the cerebrospinal fluid in the ventricles (Pollay and Davson, unpublished) similar to the active process described by Pappenheimer,

HEISEY and JORDAN (1961) for Diodrast and phenol red. Thus the higher the concentration of thiocyanate in the blood the higher will be the concentration in the cerebrospinal fluid at the steady state, and thus the *smaller* will be the "sink-action" of this fluid on the extracellular space of the nervous tissue. Hence the larger will be the measured thiocyanate-space. Thus the *upper* value of 17% will be nearer the true thiocyanate-space than the lower value of 4% somewhat arbitrarily chosen by STREICHER. Since there is no obvious reason why the thiocyanate ion should not distribute itself between extracellular space and cells in the same way as the chloride ion, it is likely that the true thiocyanate-space will be equal to the chloride-space, i.e. considerably higher than the true extracellular space.

VAN HARREVELD and SCHADE (1960) measured the electrical impedance of brain tissue and showed that it was consistent with the presence of some 27—35% of low-resistance material, i.e. extracellular fluid; this would bring the extracellular space close to the chloride-space of brain (MANERY and HASTINGS 1939; DAVSON 1955; WOODBURY et al. 1956). This figure was obtained by substitution of the measured electrical parameters in an equation derived by MAXWELL for the impedance of a suspension of particles and, according to HORSTMANN and MEVES (1959), its applicability to a tissue is not completely justified, so that a small change in some of the parameters could easily reduce the computed value to some 8%.

Because of the blood-brain barrier and the "sink-action" of the cerebro-spinal fluid, the best estimates will eventually be based on studies on excised tissue; but until the tissue can be maintained strictly normal in an artificial medium the measured values will be subject to some uncertainty. To achieve this condition the tissue should be cut very thin so that oxygenation is adequate, but this involves damage to a large proportion of the cells by the cutting knife. When thick pieces of cerebral hemisphere are used, involving a minimum of cutting, defective oxygenation must cause water-shifts between cellular and extracellular compartments, but the precise nature of the changes that take place under these conditions is not entirely clear, since PAPPIUS and ELLIOTT (1956) describe a swelling that seems to be purely extracellular, whereas electron-microscopical studies suggest that it is entirely intracellular under the same conditions. The studies of PAPPIUS and ELLIOTT (1956) suggest that the sucrose- and thiocyanate-spaces of rat cortex are of the same order of magnitude as the chloride-space.

From a functional point of view, then, it is difficult to escape the conclusion that the brain and cord contain an appreciable space in which diffusion of sucrose and other extracellular tags is relatively unrestricted; the *in vivo* study, in which the tag was presented by barbotage, probably gives the best value, namely about 12% of the total tissue volume, so that a considerable proportion of the chloride must be intracellular, and this is consistent with the finding of DAVSON and SPAZIANI (1959) that when brain slices are placed in a nitrate medium the chloride diffuses out as though it were contained in two compartments, some 30% being removed much more slowly than the rest, which diffused out as though through an aqueous solution. The problem remains, therefore, to reconcile the purely morphological studies, demanding a maximum extracellular space of some 4—5%, with the functional studies indicating a minimum in the region of 12%. The tendency now is to achieve this reconciliation by identifying the glial cells with the functional extracellular space of the physiologist. The evidence bearing on this is derived from electron-microscopical studies of the morphological changes in the glial cells observed when the brain suffers alterations in its volume.

4. Cerebral oedema and cellular hydration. Under various experimental and pathological conditions the volume of the brain may increase or decrease; thus when the volume of an excised normal brain is measured it is found to be some 10—12% less than the total cranial capacity; as a result of concussion, for example, the deficiency in volume falls to 6% and this seems to be due to an uptake of fluid by the parenchyma rather than to any vascular or ventricular engorgement (WHITE, BROOKS, GOLDTHWAIT and ADAMS 1943). The problem as to the extent to which an increase in brain volume is due to an increase in extracellular fluid as, for example, would be observed in lymphatic obstruction in a limb, or is due to an increase in intracellular water as typically observed in a tissue in which the normal pumping mechanisms for extruding sodium have failed, has not been resolved, and it is best to envisage both possibilities. Recent studies with the electron microscope tend to emphasize the importance of cellular swelling in oedema, and cellular contraction in experimental shrinkage of the brain by hypertonic solutions. For example, GERSCHENFELD, WALD, ZADUNAISKY and DE ROBERTIS (1959) maintained brain slices in artificial media, under which conditions, of course, they swell; according to them the swelling was *intracellular*, in the sense that the astrocytes increased greatly in volume although the neurones were said actually to be compressed by the expanding astrocytes. There was no appearance of extra-cellular spaces under these conditions. The surprising feature of these results is not that the swelling was intracellular — this is rather what would be expected of a tissue under anoxic conditions[1], the loss of K^+ being more than compensated by a gain of Na^+ — the surprising feature is the failure of the neurones to swell, and this may be an expression of the greater sensitivity of the astrocytes to anoxia.

What is indeed a striking feature of this work and deserves further investigation is the apparent immunity of the brain from conditions that

---

[1] Thus VAN HARREVELD and OCHS (1956) observed a large and rapid increase in brain impedance after circulatory arrest, and this could best be interpreted on the assumption of a rapid movement of extracellular fluid into the cells.

cause oedema in the rest of the body, e.g. the water intoxication due to ingestion of large quantities of water by pitressin-treated animals caused no change in water content of the brain; again, expansion of the total fluid volume of the animal to give a 30—40 % increase in body weight, by intra-peritoneal injections of saline, failed to change the water content or micro-scopical appearance of the brain.

Another way to produce oedema in the rat is to treat it with triethyl tin compounds (MAGEE, STONER and BARNES 1957). The oedema seems to be confined, by ordinary light-microscopical inspection, to the white matter and was described by MAGEE et al. as "interstitial". Electron-microscopical studies of TORACK, TERRY and ZIMMERMAN (1960) suggest that the oedema is similar to that taking place in cold injury (TORACK et al. 1959) and character-ized by enlarged glial perivascular cell processes. As MAGEE et al. had found, the oedema was not associated with passage of trypan blue from the blood into the oedematous tissue, i.e. there was no "breakdown of the barrier".

By changing the osmotic pressure of the blood WEED and MCKIBBEN (1919) were able to demonstrate large changes in the water content of the brain; this was considered to be due to changes in extracellular space so that the so-called Virchow-Robin spaces, surrounding the blood vessels, were expanded by hypotonic solutions and shrunk by hypertonic ones by the difference of osmotic pressure operating across the capillary endothelium. According to LUSE and HARRIS (1960, 1961) however, the changes in volume are experienced by the glial cells (oligodendroglia according to Luse's nomen-clature but presumably astrocytes according to other workers such as SCHULTZ, MAYNARD and PEASE (1956); see, p. 33). Thus the capillaries are stated by LUSE (1956) to be completely invested by oligodendroglial and astrocytic processes; in the oedematous condition the oligodendrocytic processes swell enormously so that the astrocytic processes are almost squeezed out and appear as tenuous strands between the swollen oligodendrocytes; in the light microscope this would give the appearance of strands crossing a swollen peri-vascular space and would thus account for WEED and MCKIBBEN's description of an enlarged perivascular or Virchow-Robin space. By injecting hypertonic solutions LUSE and HARRIS (1960) observed that the consequent dehydration was accompanied by shrinkage of the oligodendrocytes; under these conditions, however, the neurones also showed a corresponding shrinkage and it seems likely that the failure to observe a swelling under conditions of hypotonicity was a result of limitations of space, the glial swelling limiting the possible neuronal swelling and obscuring from observation such swelling as did occur.

In general, then, the conclusion that changes in volume of the brain are frequently to be ascribed to changes in volume of the cellular compartment, in particular the glial, cannot be seriously contested, but the argument from this, that all exchanges between blood and neurones take place through the medium of the glia, does not necessarily follow. Thus, if we accept the presence of a very exiguous extracellular fluid, the osmotic effects would be exerted primarily through this across the blood-brain barrier, so that the order of events on making the plasma hypertonic might be: loss of water from cerebrospinal fluid to plasma in choroid plexuses; loss of water from extracellular fluid to plasma; the small volume of extracellular fluid would mean that only a small decrease in its volume would cause the osmotic pressure to rise to that in the plasma; as a result the cells would give up water to the extracellular fluid reducing its osmolarity so that this would immediately be transferred to the plasma. Thus, so long as we admit the presence of a barrier across the capillary capable of producing a difference of osmotic pressure, the cellular shrinkage observed morphologically follows logically, so that the only problem not adequately explained is the apparent preferential participation of the glia in the osmotic exchanges, but this may be more apparent than real, as indicated above. Under abnormal conditions the water uptake by the brain is also usually cellular but this is to be expected when the abnormality reduces the efficacy of the electrolyte pumps that normally maintain the electrolyte composition, and secondarily the water content, of the cells constant. The fact that the glial cells show the effects most strongly could be due to their greater susceptibility to anoxia and toxic influences.

This greater susceptibility could be due to their containing mainly sodium as their cation instead of potassium, as postulated to explain the high sodium space of brain (p. 25). A "sodium cell" only has to pump out sodium to maintain its normal water-content, which tends to increase because of the Donnan-swelling due to the presence of colloids in the cell. Depletion of metabolic energy, through anoxia or toxic influences, puts the pump out of action and the immediate effect is to allow sodium to enter and then water. With a potassium cell, the extrusion of sodium is associated with accumulation of potassium, so that failure of the sodium pump leads to loss of potassium and gain of sodium; it is only because gain of sodium preponderates over loss of potassium that the cell swells. Hence the swelling of the sodium cell may be expected to be more rapid.

5. Other studies on brain oedema. We may recapitulate here, very briefly, some recent studies on brain swelling. LEE and OLSZEWSKI (1959) studied the behaviour of labelled plasma albumin during heat-damage to the brain; the damaged tissue showed intense blackening, indicating escape of radioactive albumin from the vascular compartment, before there was any histological appearance of oedema, i.e. before cellular changes took place. This suggests that the first reaction to heat is an increased capillary permeability to proteins similar to that in other parts of the body. According to KLATZO, PIRAUX and LASKOWSKI (1958) the oedema due to local cooling of the pia is manifest as astrocytic swelling, which appears well before the breakdown of the blood-brain barrier to fluorescein. ISHII, HAYNER, KELLY and EVANS (1959) caused oedema by compression of the brain and found the barrier to diiodofluorescein reduced; they considered that the oedema was interstitial, but only on the basis of light-microscopical appearances. KIYOTA [1959 (a), (b)] has extracted the proteins from homogenates of normal and oedematous brains, the last being autopsy material of various pathology. In some cases the changes in protein content indicated the passage of plasma albumin into the tissue to give what he called "complex oedema", by contrast with "simple oedema" when the only change measurable was in the water content, the extracted proteins being apparently normal in composition. STERN (1959) has measured brain volume after intravenous injections of distilled water; according to him the increase in brain bulk so frequently described is more apparent than real; moreover it was only the grey matter that showed an increase in water-content (80.3 ± 0.4 to 81.0 ± 0.5), the white matter showing an actual decrease (71.6 ± 1.4 to 70.1 ± 1.0). In striking contrast to those studies that have shown brain oedema to be due to an intracellular swelling is the study of MAGEE, STONER and BARNES (1957) using triethyl tin compounds; here the swelling was confined to the white matter and seemed to be purely interstitial.

Asphyxia causes a decrease in the volume of extracellular fluid as determined by the impedance of the tissue (VAN HARREVELD and OCHS 1956); histochemical studies (VAN HARREVELD 1961) show that this is associated with the migration of chloride into dendrites of neurones and glial cells of the cerebellar cortex; in the cerebral cortex, on the other hand, the migration seems to be entirely into the apical dendrites of pyramidal cells. When electrolytes are washed out of the brain *in vivo* by continuous infusion of glucose solution the impedance rises, indicating that this parameter is, indeed, a measure of extracellular electrolytes (VAN HARREVELD, HOOPER and CUSICK 1961).

Post mortem the brain swells rapidly, e.g. in 15 min a guinea pig brain increased from an average value at death of 4 g to an average value of 4.15 g; this must be due to a post-mortem absorption of cerebrospinal fluid, presumably by the cells (EDSTRÖM and ESSEX 1955).

It seems quite obvious that a great deal of careful experimentation on brain oedema is required. In particular, it is important that the chemical and morphological changes taking place should be studied in the same tissue; where the blood-brain barrier is investigated too, as in traumatic influences, the use of dyes should be avoided since these only indicate tissue staining and may be just as much a measure of cellular damage as an alteration of the blood-brain barrier. Finally, the oedemas due to different experimental and pathological influences should be carefully kept separate in any discussion.

D. Further experimental evidence on the nature of the blood-brain barrier

1. Morphological studies. The existence of a barrier, as revealed classically by the failure of trypan blue to stain the brain, was early attributed to the presence of astrocytic feet that, in the light microscope, apparently enveloped the blood capillary, so that a substance in the latter had to cross not only the endothelial lining of this vessel but also the protoplasmic sheath constituted by glial end-feet. The alternative suggestion, that the capillary endothelium was fundamentally different, was also considered. The high resolution permitted by the electron microscope applied to thin sections of brain has increased our knowledge of the relationships between capillaries, glia and neurones, but unfortunately not to the point where we can state unequivocally the morphological correlate of the blood-brain barrier.

It will be beyond the scope of this review to discuss the morphology of the glial cells in full, and the interested reader is referred to a recent symposium (WINDLE 1958); here we are concerned with their relation to the capillaries on the one hand and to the neurones on the other. According to the study of SCHULTZ, MAYNARD and PEASE (1957), and MAYNARD, SCHULTZ and PEASE (1957), the astrocytes are characterized by a watery cytoplasm with processes that ensheath the capillary endothelium; by contrast, the oligodendroglia are small, with denser cytoplasm appearing as a narrow rim round the nucleus, and few processes. They take up a "satellite" position in relation to neurones. The microglia are small motile phagocytic elements. The capillary consists of an endothelial layer of flat cells resting on a basement membrane which faces outwards, away from the lumen. This basement membrane is shared by the overlying astrocytic protoplasm to produce a stituation similar to that observed in the kidney glomerulus where an epithelial layer covers the endothelial layer of the capillary. The total gap between endothelial cell and astrocytic process is some 300—500 Å and is made up of a central zone filled with dense homogeneous material (*lamina densa*) and between this and the cellular layers a less dense zone (*lamina rara*). Thus the blood-brain barrier, viewed electron-microscopically, consists of the endothelial cells, the dense homogeneous basement membrane material, and the astrocytic processes. MAYNARD et al. emphasize that the basement membrane is no different in appearance from that observed in capillaries of other parts of the body, so that the special features of the blood-brain barrier are not to be attributed to this. As the present author has emphasized earlier (DAVSON 1956), the astrocytic layer, in order to act as an effective barrier, must be continuous over the whole capillary; if appreciable gaps between nearby processes occur, the effectiveness of the processes as a barrier will be reduced to negligible proportions (thus the high degree of permeability shown by muscle capillaries can be ascribed to discontinuities between endothelial cells amounting to less than one per cent of the whole surface, PAPPENHEIMER 1953). According to MAYNARD, SCHULTZ and PEASE (1957) the astrocytic processes form an incomplete layer, representing only some 85 % of the total surface; on the other hand DEMPSEY and WISLOCKI (1955), LUSE (1956), and LUSE and HARRIS (1960) are emphatic in stating that the capillaries are completely invested by glial processes. (In the retina, which is an outgrowth of the central nervous system, the capillaries are said to be completely invested by glial protoplasm, KISSEN and BLOODWORTH 1961.) Until this morphological contradiction can be resolved, therefore, the question as to whether the barrier is at the endothelium, the basement membrane, or the glial covering, must remain open.

BENNETT, LUFT and HAMPTON (1959) have reviewed the types of capillary, as revealed by electron microscopy, in different tissues and have proposed a convenient notation to indicate their characteristic features. The capillaries of the cerebral tissue have a continuous basement membrane, a feature they share with muscle; they have no fenestrations in the endothelial cells, nor yet large intercellular gaps, and finally they are well, if not completely, covered by a pericapillary investment. They are similar to muscle capillaries in all these respects with the exception of pericapillary investment, which is less than 50% of the surface here. Capillaries of the intestine and choroid plexus contain intracellular fenestrations, whilst the liver and spleen sinuses contain large intercellular fenestrations. Attempts to locate the site of the barrier by electron-microscopical identification of material that has migrated out have been made by DEMPSEY and WISLOCKI (1955) and VAN BREEMAN and CLEMENTE (1955), who-fed silver nitrate to rats over many months and finally examined their brains. The deposits of silver were prominent in the connective tissue surrounding the capillaries in regions where there is no apparent barrier, i.e. the area postrema, infundibulum, etc. By contrast, in regions where there is a barrier and no collagenous connective tissue, there was little or no silver deposited so that it was concluded that the barrier was the endothelial cells or their glial investment. Essentially, of course, these authors were describing the presence of argyrophilic areas and not necessarily areas that permitted access of silver. In one respect the capillaries of the nervous tissue seem to be different, and that is in regard to the presence of a high concentration of ATPase in their walls (TORACK, BESEN and BECKER 1961).

The capillaries of the choroid plexus do not, themselves, exhibit a barrier, which is rather the property of the epithelial layer covering them; it is therefore of some interest to compare the electron-microscopical appearances of the capillaries of the choroid plexuses and those of the brain parenchyma. According to the study of MAXWELL and PEASE (1956) and WISLOCKI and LADMAN (1958), the endothelium belongs to the class seen in the peritubular capillaries, the cells having definite, although sparse, intracellular fenestrations. The endothelium rests on a thin dense basement membrane. Thus the difference between plexus capillary and parenchymatous capillary resides in the presence of gaps in the endothelial cell membranes, and this could easily account for the absence of a barrier between plexus capillary and the surrounding connective tissue, the blood-cerebrospinal fluid barrier being a function of the choroidal epithelium. Since muscle capillaries have no endothelial intracellular fenestrations, however, we cannot attribute the presence of a blood-brain barrier to the mere absence of fenestrations.

Glial cells as intermediaries in nutrition. The close investment of the cerebral capillaries by processes from astrocytes has naturally led to the suggestion that they may act as intermediaries in the nutrition of the neurones; thus, in making a suggestion of this sort, SCHULTZ et al. (1957) state that they were impressed with the manner in which astrocytes and their cytoplasm filled in the interstices in the neuropil, their total volume constituting a notable fraction of the bulk of the grey matter. Capillaries, they noted, were rather sparsely distributed, so that metabolites might have to travel considerable distances, and with no large intercellular pathways such as those in muscle. The anatomical relationships thus strongly suggested to them that the astrocytes might serve as the principal transport system, a view taken up by FARQUHAR and HARTMANN (1957) on the basis of their electron-microscopical study. In this connection the function of the oligodendrocytes cannot be ignored; according to PALAY'S (1958) electron-microscopical study they constitute some 50% of the glia (astrocytes 40% and microglia 6—10%), and they would seem to be closely associated with the neurone (especially as revealed by the process of satellitosis). Their functions have been variously described as to insulate synapses, to act as glands of secretion, or to synthesize myelin[1]; because of their closer association with the neurones and more remote association with the vascular tree it has been suggested (SCHEIBEL and SCHEIBEL 1958) that they are linked in a bucket-carrying mechanism whereby metabolites pass from capillary to astrocyte, to oligodendroglia, to neurone. According to CAMMERMEYER [1960 (a), (b)], however, on the basis of light microscopy of serial sections, the oligodendroglia, because of their constant association with blood vessels, especially in the region of bifurcation, were able to control the diameters of these vessels, presumably by virtue of their contractile powers. They would thus serve as a local control mechanism in the nutrition of the neurones. CAMMERMEYER considered that neurones definitely made contact with capillaries so that there was a direct pathway of metabolites from capillary to neurone as well as an indirect one by way of the astrocytes. There is reason to believe that the oligodendroglia are capable of contractile activity, since CANTI, BLAND and RUSSELL (1937), LUMSDEN and POMERAT (1951) and BERG and KALLÉN (1959) have described pulsatile activity in tissue-cultured cells, an activity that can be modified by serotonin (WOOLLEY and SHAW 1957). The pulsatile activity would certainly contribute to the passage of material from capillary to neurone, which otherwise must be a very inefficient mode of transport, involving as it would the crossing of several cell barriers.

Changes in the amount of Nissl substance of spinal neurones have been shown by KULENKAMPFF (1952) to be accompanied by the aggregation of satellite neuroglia around them; moreover, the size of the nucleus of a satellite glial cell seems to increase with muscular activity on the part of the animal. NIESSING (1950) has correlated changes in the number of processes belonging to astrocytes with narcosis.

In a recent study, KLATZO and MIQUEL (1960) have shown that fluorescein-labelled protein is taken up by microglia and astrocytes in tissue culture; according to them this indicates pinocytosis. When the brain is damaged by cold it is the astrocytes and oligodendroglia that show accumulation of the intravenously injected protein.

According to OKSCHE (1958) in primitive forms such as the frog the ependymal cells carried glycogen to the neurones; as the brain became thicker this function was taken over by the vascular system and astrocytes.

2. Functional studies. That the endothelial cells of the cerebral capillaries are different from those in other parts of the body is suggested by RODRIGUEZ' (1955) study with acridine dyes which, when given intravenously, stain the nuclei of all cells in the body except those of the central nervous system, including the nuclei of the capillary endothelial cells. When the dye (proflavine hydrochloride) was introduced into the cerebrospinal fluid the nuclei of all the cells of the central nervous system, including the endothelial nuclei, were

---

[1] The electron-microscopical studies of DE ROBERTIS, GERSCHENFELD and WALD (1958) provide convincing evidence for the involvement of the oligodendroglia in laying down myelin in the central nervous system; hence to attribute an additional role to these cells seems an act of supererogation unless the evidence is strong. The rapid nucleic acid and protein turnover in the oligodendroglia may well be related to this aspect of their activity (KOENIG 1958).

now stained. Thus the endothelial cells appear to have an asymmetry, their luminal sides, facing the blood, preventing the passage of the dye into their cytoplasm and nuclei, whilst the sides facing the tissue cells seems to permit access to the endothelial cell.

This asymmetry may, however, be only apparent, since a large proportion of the dye, when circulating in the plasma, may be bound to plasma proteins; hence the effective available concentration must be much greater when the dye is introduced directly into the cerebrospinal fluid. This point was first made by TSCHIRGI (1950) in respect of trypan blue.

Another line of evidence exhibiting a barrier between blood and the nervous tissue is provided by a recent study by BAKAY, BALLANTINE and BELL (1959). When nervous tissue is damaged by ultrasonic irradiation, the uptake of intravenous $^{32}PO_4$ is more rapid and extensive than normal (BAKAY, HUETER, BALLANTINE and SOSA 1956); this is the characteristic breakdown of the blood-brain barrier, but it could be argued that it was due to an abnormality of the cells of the damaged parenchyma which took up more phosphate as a result. However, if the $^{32}PO_4$ is presented by way of the cerebrospinal fluid, the uptake by the damaged parts of the tissue is normal. Thus, in the normal tissue, phosphate passes slowly from the blood; in the damaged tissue it passes more rapidly from the blood, but not from the cerebrospinal fluid. Hence the damage has indeed broken down a pre-existing barrier. An electron-microscopical study of the changes in the capillary-glial relationships resulting from damage would be of great interest. In the light microscope, breakdown of the barrier to trypan blue seems to be associated with degenerative changes in the endothelial, rather than the glial, cells (GRÖNTOFT 1954).

The blood-brain barrier was originally demonstrated by Goldmann's so-called first and second experiments; according to the first experiment, intravenous trypan blue failed to stain the brain whilst, according to the second experiment, trypan blue injected into the cerebrospinal fluid did stain the tissue. Subsequent studies have shown that a variety of substances, when injected into the cerebrospinal fluid, find their way rapidly through the parenchyma, although when injected intravenously they fail to do so. With ^{24}Na, penetration from blood to nervous tissue is fairly rapid by comparison with, say, iodide and thiocyanate, the half-life for equilibration being of the order of 2 hr (WANG 1948; SWEET et al. 1949; TUBIANA, BENDA and CONSTANS 1951; DAVSON 1955; OLSEN and RUDOLPH 1955). Injected into the cerebrospinal fluid, it also leaves fairly rapidly, but the assessment of the relative ease with which this isotope crosses the blood-brain barrier, on the one hand, and the cerebrospinal fluid-brain barrier on the other, is not easy because we do not know the relative areas of tissue exposed to the isotope in the two instances. According to BAKAY (1960), under comparable conditions of concentration-difference, the flux of ^{24}Na is some three times more rapid in the direction: cerebrospinal fluid to brain, than when the movement is from brain to cerebrospinal fluid. According to COULTER (1958), the area of the capillaries in the cat's brain is 1000—1300 cm^2, whilst the area of the surface of the brain is 100 cm^2; hence, in spite of the fact that the available area for diffusion is ten times greater, passage from blood to brain (blood-brain barrier) is three times slower than passage from cerebrospinal fluid to brain; the blood-brain barrier is thus some 30 times greater than the cerebrospinal fluid-brain barrier, so far as ^{24}Na is concerned[1].

3. Ontogeny of the blood-brain barrier. It has been accepted for a long time, on the basis of the early studies of BEHNSEN (1927) and STERN and PEYROT (1927), that in many species, e.g. mouse, rabbit, cat, the blood-brain barrier is immature at birth or in the foetus, requiring several days post natum to acquire its ability to exclude trypan blue. According to BAKAY (1953) this applies also to the uptake of intravenously injected ^{32}PO$_4$ in the rabbit[2]. Recent studies have cast some doubt on these claims. Thus GRÖNTOFT (1954) profited by BROMAN'S (1950) observation that the barrier remains intact in the dead animal for at least 12 hours, so that pre-viable foetuses, and prematurely deceased infants, could be studied by a perfusion technique. Human foetuses showed the presence of a barrier to trypan blue at 5—30 cm length, and in newborn infants the barrier was as complete as in the adult. In guinea pigs of 1—3 days' age there was a good barrier, and in rabbits weighing 4 to 22 g there was no barrier. The perfection of the barrier at birth in the guinea pig had already been demonstrated by STERN and PEYROT and was interpreted by them as a sign of the more complete development of this animal at birth than that of the rabbit and cat. In a more extensive study of the foetal rat by GRAZER and CLEMENTE (1957), who made their injections through the wall of the maternal uterus, found no penetration of trypan blue into the central nervous system, even in animals less than twelve days after their conception. AGAIN, MILLEN and HESS (1958) found that the barrier to trypan blue in 2—8 day-old rats was just as effective as in adults. In contrast with these results, however, there is the claim of LAJTHA (1957) that the exchanges of ^{35}Cl and thiocyanate that take place between

---

[1] Another example, illustrating barrier function, is given by the amino acid methionine; administration of ^{35}S-labelled methionine cisternally leads to a rapid incorporation into the brain proteins by comparison with a slow incorporation when the amino acid is given intravenously (GAITONDE and RICHTER 1956). AGAIN, HIMWICH, PETERSEN and ALLEN (1957) found that the injection of glutamate into 24 hour-old rats caused convulsions, whereas these were rare if the animals were 17 days old. DOBBING (1961) has emphasized the care that must be exercised in interpreting this sort of experiment, however, since the results may simply reflect different metabolic activity rather than different barrier function.

[2] However, GRUHN (1957) states that in the newborn kitten the rate of uptake of ^{32}PO$_4$ by the cerebrospinal fluid is the same as in the mother; moreover, uptake by the brain of the foetus was similar to that of the mother.

blood and brain during one hour after parenteral injection are smaller in embryonic and developing chicks and rats than in older animals, but the figures that they present are not absolutely convincing on this point.

Since the demonstration of clear-cut differences in the barrier phenomenon according to the age of the animal would be convincing evidence of the *existence* of a barrier, and would also provide a basis for the experimental study of the morphological features that determine the barrier, it is important that the truth or otherwise of the assertion that the barrier is incomplete in foetal and newborn animals should be investigated further by the use of quantitative techniques; thus the mere statement that the brain stains or fails to stain with trypan blue is not sufficient evidence respecting the permeability of the barrier as such; nor yet is the observation that $^{32}PO_4$ accumulates more rapidly in the nervous tissue after intravenous injection (BAKAY 1953) since its incorporation is a function both of the barrier and the metabolism of the tissue. This is also true of an amino acid like lysine which, according to LAJTHA (1958), is taken up to a greater extent by the brains of young mice than older ones.

4. Metabolic aspects of the blood-brain barrier. This brings us to an aspect of the blood-brain barrier that has been emphasized recently by DOBBING (1961), namely the involvement of many substances in the metabolism of the nervous tissue. Thus, the failure to observe an appreciable uptake of certain substances by the brain when these have been administered intravenously can obviously be due to the circumstance that they are involved in metabolic activity as soon as they enter the brain parenchyma and lose their identity. From the present point of view this aspect is of interest, therefore, to the extent that this involvement in metabolism may complicate the measurement and interpretation of "barrier phenomena". The outstanding studies in this respect are those of LAJTHA and his colleagues on the amino acids (LAJTHA, FURST, GERSTEIN and WAELSCH 1957; LAJTHA, BERL and WAELSCH 1959; LAJTHA 1959; LAJTHA and MELA 1961). In this work a distinction between the exchange of intravenously injected amino acid with the pool of free amino acids in the brain — the true barrier phenomenon — and the subsequent incorporation into proteins, was maintained. As Fig. 7 shows, for the case of leucine, the increase in specific activity of the free leucine in the brain occurs at the same rate as that in muscle, suggesting that the passage out of the blood stream is not the determining factor, but rather the passage from the extra-cellular space into the cells of the respective tissues. From Fig. 7 it is possible to compute a turnover-constant of 0.04 min^{-1}, of the same order as that for the relatively lipid-soluble propyl thiourea. This relatively rapid rate of entry into the cells is presumably an example of facilitated transfer.

In some instances there seems little doubt that passage of the barrier is the rate-limiting step in the incorporation of material in the blood by the brain; thus with $^{32}PO_4$ STREICHER (1956) studied its incorporation into various fractions (inorganic, phospholipid, phosphoprotein, total nucleic acid, etc.) of the brain phosphorus and found that the rates were characteristically different and in the order one would expect from the known sequence of metabolic events; in other words, the entry of phosphate into the brain from the blood was *not* the rate-limiting factor because, if it had been, the rates of incorporation into the different fractions would have been about the same. At later times after the intravenous injection of phosphate, however, rates of incorporation into many of the fractions became about the same, so now the barrier had become the rate-limiting factor, and this was because, at the later times, the concentration in the plasma had fallen so low that the concentration gradient across the barrier had become small.

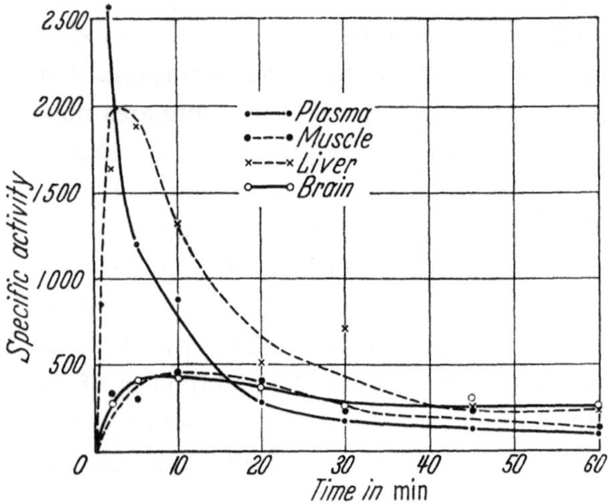

Fig. 7. Specific activity of free lysine in mouse organs after administration of C[14]-lysine. (From LAJTHA, 1957)

5. Sugars. The failure of the perfused brain to utilize the glucose in the perfusion fluid unless a liver extract was added (GEIGER, MAGNES, TAYLOR and VERALLI 1954) has been further investigated. It has been shown that addition of cytidine or uridine enables the brain to take up glucose (GEIGER and YAMASAKI 1956) and remain viable for up to four hours of perfusion, compared with only one hour in their absence; their function seems to be to correct the faulty carbohydrate metabolism that results from the failure to take up glucose; uridine preventing the depletion of galactosides, and cytidine the depletion of phospholipids that otherwise occur. 5-Hydroxytryptamine (serotonin) added to the perfusion fluid increased the uptake of sugar a little but did not improve the condition of the brain (MAGNES and HESTRIN-LERNER 1959). For a review of this aspect, see GEIGER (1958)[1].

6. Urea. When the osmolarity of the blood is raised, the brain shrinks as a result of the withdrawal of water; particularly efficacious in causing this

---

[1] It is not conceivable that the brain and cord should depend entirely for their glucose supply on the circulating cerebrospinal fluid; in fact WOLFF and TSCHIRGI (1956) found that perfusing the subarachnoid space of insulin-treated cats with glucose solutions failed to prolong the reflex activity of the cord. GREIG and GIBBONS (1959) have shown that the incorporation of ^{14}C into the brain after administration of ^{14}C-glucose is delayed by chlorpromazine and psychomimetics like mescaline, but this could mean anything, and is more likely to represent an action of the drugs on carbohydrate metabolism than on the blood-brain barrier.

shrinkage is intravenous or oral urea (FREMONT-SMITH and FORBES 1927; SMYTHE, SMYTHE and SETTLAGE 1950; JAVID and SETTLAGE 1956; STUBBS and PENNYBACKER 1960). The effectiveness of urea is due to two factors, namely the slow rate of excretion by the kidneys and the low permeability of the blood-brain barrier to this solute[1]. The blood-brain and blood-cerebro-spinal fluid barriers to urea have been studied by SCHOOLAR, BARLOW and ROTH (1960), by BRADBURY and COXON (personal communication) and KLEEMAN, DAVSON and LEVIN (1962); all authors are agreed that the white matter equilibrates more slowly than the grey, presumably because of the greater vascularity of the latter; penetration into the cerebrospinal fluid is rather slower than into the water of the grey matter, hence the removal of water by plasma made hypertonic with urea is probably not confined to the brain, but is shared by the cerebrospinal fluid. This would be obscured, how-ever, by the reduced drainage of cerebrospinal fluid immediately following the fall in intracranial pressure caused by the urea, so that the volume of cerebro-spinal fluid might well rise in spite of some degree of concentration. It is well established that the concentration of urea is lower in the cerebrospinal fluid than in the plasma under normal steady-state conditions, the value of R_{csf} being given variously as 0.58, 0.64, and 0.82. It might be expected that the water of the brain-tissue would have a concentration between that in the cerebrospinal fluid and plasma, but careful study of steady-state distributions between plasma on the one hand and cerebrospinal fluid, white matter and grey matter on the other, indicate that the position is highly complex. Grey matter actually has a higher concentration of urea in its water than that in the plasma (BRADBURY and COXON; KLEEMAN et al. 1962) whilst the concen-tration in the water of the white matter is about equal to that in the plasma. It is of great interest, therefore, to consider how the nervous tissue can maintain a concentration of urea not only higher than that in the plasma but also considerably greater than that in the cerebrospinal fluid. Urea can exchange fairly rapidly across the cerebrospinal fluid-brain barrier (DAVSON, KLEEMAN and LEVIN 1962). There is some evidence that urea may be synthesized by the brain (SPORN, DINGMAN, DEFALCO and DAVIES 1959) and this might account for its normally high concentration in grey matter, but the failure of this to diffuse into the cerebrospinal fluid, giving high concentrations in the subarachnoid fluid by comparison with the ventricular (cf. p. 45), is difficult to explain unless we assume that some cells, perhaps the glia, actively main-tain a low concentration of urea within them so that diffusion between cerebro-spinal fluid and nervous tissue is relatively negligible.

---

[1] In hypothermia the excretion of urea is reduced, so that hypertonic urea, adminis-tered intravenously, is very effective in lowering intracranial volume (BERING and AVMAN 1960).

III. Relationship between cerebrospinal fluid and nervous tissue

The fundamental fact that the cerebrospinal fluid and nervous tissue of brain and cord are in close diffusional relationships has been well recognized since the studies of Goldmann and of Stern, but quantitative measurements of these relationships are of relatively recent date. In general, if a substance is introduced into the cerebrospinal fluid, it can leave by flow through the arachnoid villi, by diffusion into the blood in the choroid plexuses (if injection is into the ventricles), and by diffusion into the nervous tissue and from there into the blood by way of the capillaries of this tissue.

In the study of these relationships, two main types of experiment have been carried out. In those of the first type, a known amount of a substance is injected into the cerebrospinal fluid and after a given time the amount escaping is computed by analysis of the remaining fluid. Quantitative comparisons between different substances are best made by including in the injected solution a standard "tag", e.g. ^{24}Na, so that the losses of all substances studied can be expressed in terms of the loss of the tag. In this way the great animal-to-animal variation can be largely compensated for. In experiments of the second type, absorption is confined to a limited region of the central nervous system by perfusing a solution through the ventriculo-subarachnoid system, as in the ventriculo-cisternal perfusion of Leusen (1948, 1950) and Royer (1950) who placed an inlet-cannula in the lateral ventricle and an outlet-cannula in the cisterna magna. As developed by Pappenheimer, Heisey and Jordan (1961) in the goat, this method allows of a precise assessment of the "clearance" of a given substance from the cerebrospinal fluid into the blood stream. By combining the analysis of inflow- and outflow-fluids with a post mortem study of the distribution of the perfused substance in the parenchyma lining the ventricles, some knowledge of the extent to which material may diffuse into the parenchyma, before being washed away by the blood stream of this tissue, may be obtained (Draškoci, Feldberg, Fleischhauer and Haranath 1960).

We may consider the more quantitative studies first. Pappenheimer et al. perfused creatinine, inulin and Diodrast from lateral ventricle to cisterna magna of the goat and showed that these substances were absorbed to different extents, the remarkable feature being the high rate of absorption of Diodrast by comparison with that of the smaller creatinine molecule; this suggested an active transport process, similar to the tubular secretion of this substance by the kidney (but in the reverse direction), and this was proved by showing that the rate of passage from cerebrospinal fluid to blood was some 15 times the rate of passage from blood to cerebrospinal fluid. Moreover, when the concentration of Diodrast in the blood was maintained at twice the average concentration in the perfusion fluid, the absorption continued against the concentration gradient. Again, the absorption could be slowed by self-inhibition, in the sense that raising the concentration of Diodrast above a few mg-% apparently saturated a carrier-mechanism, and finally it could be inhibited by competition with p-aminohippurate (PAH). The remarkable finding was that it was apparently only the choroid plexus of the IVth ventricle that was responsible for the active absorption. This was shown by a stop-flow technique; after stopping flow and then resuming it, and taking small successive samples of effluent, it was found that maximal absorption had occurred in the earliest samples, i.e. those that were directly from the cisterna magna and IVth ventricle. This description of an active absorption by the choroid plexus had been anticipated by tissue culture studies of CAMERON (1953) and LUMSDEN (1958).

In the cat, using the same technique and quantitative approach, DAVSON, KLEEMAN and LEVIN (1962b) studied the absorption of inulin, ^{24}Na, ^{14}C-urea, PAH and ethyl alcohol presented in the same perfusion fluid. When the inlet-cannula was in the lateral ventricle, the absorption of PAH was more rapid than that of ^{24}Na, and this strongly indicated the intervention of an active process, since when these two substances are presented in the blood the penetration of PAH into the cerebrospinal fluid is negligible by comparison with that of ^{24}Na (DAVSON 1955; DAVSON and SPAZIANI 1959). As we might expect, absorption of the highly lipid-soluble ethyl alcohol was very rapid, so that the effluent never contained a measurable concentration of this substance. When the inlet-cannula was placed in the IIIrd ventricle, absorption was very much reduced, so that the fall in concentration of PAH, inulin, ^{24}Na, and ^{14}C-urea was due mainly to the mixing with newly formed fluid, whilst now with the highly lipid-soluble ethyl alcohol measurable concentrations were found in the effluent. These results emphasize the importance of the lateral ventricles for absorption, by contrast with those of PAPPENHEIMER et al. on the goat where the IVth ventricle was the important site[1].

Less quantitative studies, involving drugs or solutions of altered p_H and bicarbonate concentration, in which the emphasis has been on the pharmacological effects of the perfusion, are those of FELDBERG and MALCOLM (1959) with tubocurarine and nor-adrenaline, and DRAŠKOCI et al. (1960) on histamine. The results demonstrate that these drugs are able to exert their influence on the brain by diffusion across the ventricular and aqueductal walls, and analysis of frozen sections cut parallel to the ventricular walls indicated passage to a depth of at least 2.5 mm, passage along the grey matter being more rapid and extensive than that along the white matter. LOESCHCKE and KOEPCHEN'S (1958) work emphasized the importance of the site of action of the drug rather than its accessibility to the tissue; they showed that changes of p_H, and various drugs such as veratrine, lobeline and novocaine, would have their characteristic effects if they were perfused into the IVth ventricle so that fluid passed up into the lateral recesses of this ventricle and gained access to structures at the base of the brain. The drugs applied to

---

[1] In the rabbit, however, active absorption seems to take place throughout the ventricular system although the major portion occurs in the lateral ventricles (POLLAY and DAVSON, unpublished).

the floor of the IVth ventricle, i.e. close to the respiratory and vasomotor centres, had no influence, so it was concluded that the sensitive sites for attack were not the centres near the floor of the IVth ventricle but elsewhere. Recently DORNER (1961) has shown that the passage of ^{42}K into the cerebrospinal fluid is concentration-dependent.

When substances are injected into the cisterna magna, we may assume that access to the active absorbing sites is limited, so that we need not expect to find evidence for an active process; nevertheless the studies of DAVSON, KLEEMAN and LEVIN [1962 (a), (b)], summarized by Table 1, show that in the rabbit absorption of p-aminohippurate (PAH) is greater than that of sucrose, and that thiocyanate and iodide both show a much higher absorption than ^{24}Na. These findings suggest an active absorption of these three substances, namely PAH, thiocyanate and iodide, and this has been confirmed by ventriculo-cisternal perfusion studies (POLLAY and DAVSON, unpublished).

Table 1. *Relative losses of injected substances, expressed as percentage of the loss of ^{24}Na that was injected in the same solution. Limits are the standard errors of the difference between the loss of ^{24}Na and that of the substance in question* (DAVSON, KLEEMAN and LENIN)

Substance	Loss	Substance	Loss
Inulin	43 ± 8	^{14}C-Urea	116 ± 3
Sucrose	62 ± 3.5	Xylose	122 ± 6
p-Aminohippurate	82 ± 2	Thiourea	135 ± 4
^{35}SO$_4$	84 ± 14	Thiocyanate	140 ± 10
Creatinine	91 ± 6	Iodide	155 ± 10
^{24}Na	100	Ethylthiourea	172 ± 6

The similarity in rates of escape of inulin and sucrose shown by Table 1 indicate a non-specific bulk drainage with little impedance to flow; this view is further substantiated by experiments of a similar type carried out by PROCKOP, SCHANKER and BRODIE (1961), who compared the escapes of inulin, sucrose and mannitol from the cerebrospinal fluid after ventricular injection. The rates were similar.

These quantitative studies give us a measure of the escape of substances from the cerebrospinal fluid, which may take place directly into the blood of the choroid plexuses and of the pial circulation, and less directly by way of the nervous tissue. That this latter route is significant, i.e. that the nervous tissue participates in exchanges with cerebrospinal fluid, is proved by many qualitative studies in which the tissue has been analysed either chemically, or by radioautography in the case of isotopically labelled material. Quantitatively, moreover, the studies with barbotage, in which a relatively constant level of sucrose or PAH was maintained in the subarachnoid fluid for some time, indicated that these substances may achieve diffusion-equilibrium with the extracellular space of the cord in one to two hours. Thus the existence of this type of exchange raises the question as to how the cerebrospinal fluid maintains a peculiar chemical composition, different in many respects from that of the blood plasma, when it is exposed to nervous tissue for a long time (the half-life of the cerebrospinal fluid, on the basis of a renewal rate of $1/_2$% per min is some $2^1/_2$ hr).

Local differences in composition

We must look upon the composition of the cerebrospinal fluid, therefore, as representing a compromise between the composition of the fluid as newly secreted in the ventricles and that of the blood plasma, the more rapid the exchanges between cerebrospinal fluid and plasma, via the nervous tissue, the nearer the compositions will approximate. On this basis, then, we must expect to find obvious inhomogeneities in the fluid, according as it is close to its origin, or close to its escape route in the arachnoid villi. Thus with K, glucose and urea the concentration of ventricular fluid should be less than that of the subarachnoid fluid, whilst with Cl, Mg and Na the reverse should be true. In fact, however, differences in concentration are not easy to demonstrate, and when found are not necessarily in accordance with prediction[1]. Thus with a small animal like the rabbit, where ventricular samples large enough for analysis are difficult to obtain, we may instead compare successive small samples withdrawn from the cisterna magna. With chloride, glucose[2] and urea, successive samples were remarkably similar in concentration (DAVSON 1958; KLEEMAN, DAVSON and LEVIN 1962). ROUGEMONT, AMES, NESBITT and HOFMANN (1960) analysed the fluid collected from the exposed choroid plexus of the lateral ventricle of the cat and compared it with that from the cisterna magna and the cisterna pericallosum. The chloride concentrations were:

Plexus fluid < Cisterna magna fluid = Cisterna pericallosum and this is not what one would expect on simple kinetic grounds, the freshly secreted fluid having the lowest instead of the highest concentration. Again, with K, we should expect the plexus fluid to be least concentrated, but it was in fact more concentrated than that drawn from the cisterna magna.

These results thus demonstrate the existence of inhomogeneities, so far as Cl and K are concerned, but it is not easy to account for them on the basis

---

[1] The small differences in calcium and magnesium concentration in ventricular, cisternal and lumbar fluids, described by HUNTER and SMITH (1960), may well be due to the small differences in protein concentration in the fluids.

[2] With glucose, it may be argued that the low concentration normally observed in cerebrospinal fluid compared with that in plasma is an artefact resulting from the employment of unspecific methods of chemical analysis. Thus COOPER and ARCHDEACON (1960), using the specific glucostat method, found, in the dog, fasted for 12—84 hours, values of the ratio: Concn. in csf/Concn. in Plasma that were usually, but not always, close to unity. The comparisons were with *venous* blood, however, whilst other studies showing a large difference between plasma and cerebrospinal fluid, e.g. DAVSON (1955), employed arterial blood plasma as the standard for comparison. It is difficult to believe that the glucose of the cerebrospinal fluid is not utilized by the adjacent cellular tissue. MARKS (1960) found as the mean of 154 normal lumbar fluids in man a concentration of 60.5 ± 0.6 compared with 61.2 ± 1.4 for ventricular fluid. Although the oxidase method of determination gives consistently lower values than the Folin-Wu method, this applies to both blood and cerebrospinal fluid, so that the values of R_{csf} were not greatly different according as one or the other method was employed.

of passive exchanges between nervous tissue and cerebrospinal fluid, and we must envisage the possibility of perhaps an active absorption of K by the walls of the ventricles or perhaps by the choroid plexuses of the IIIrd and/or IVth ventricles. With bromide, again, there are indeed inhomogeneities, the concentration in the ventricular fluid being the least and that in the lumbar fluid greatest (Hunter, Smith and Taylor 1954; Bourdillon et al. 1957). Here we have the choice between assuming that the freshly secreted fluid has a low concentration of bromide, so that the concentration in subarachnoid fluid rises as a result of diffusion from nervous tissue and plasma; or we may assume that, as secreted, the concentration is relatively high but that active absorptive processes in the ventricular walls or in the choroid plexuses of the IIIrd and/or IVth ventricles lead to a reduced concentration, which then rises as the fluid flows on into the subarachnoid spaces (Lloyd and Taylor 1959).

Thus, without quoting any further work, it is apparent that the situation is at present confused in regard to the modifications undergone by the fluid from the moment of its secretion to the point at which it leaves the ventricular system, and then to the point at which it is drained into the arachnoid villi. Obviously a great deal of experimentation is needed on this aspect, but whatever the results we are left with the likelihood that, in respect to chloride and urea and possibly other substances, the concentration in the cerebrospinal fluid remains considerably different from that in the plasma with which it is capable of establishing diffusional relationships, both in the nervous tissue and in the pia. This brings us back to the problem of the extracellular space of brain and cord; if there is, indeed, an extracellular fluid, we may envisage two main possibilities. First, this fluid may be secreted by the capillaries with a composition similar to that of cerebrospinal fluid; in this way we could explain the relative homogeneity of the cerebrospinal fluid. Second, the fluid is a simple filtrate from plasma but is sufficiently small in volume, i.e. the space is sufficiently small, so that it is unable to influence the cerebrospinal fluid concentration appreciably. This hypothesis is not sufficient alone, however; we must also assume a barrier between extracellular fluid and blood, otherwise the rapid exchanges between blood and this exiguous extracellular fluid would ensure that it was indeed capable of modifying the composition of the cerebrospinal fluid. Thus a blood-brain barrier must be invoked even by those who deny the existence of an appreciable extracellular fluid[1].

---

[1] Feldberg and Fleischhauer (1960) observed that the passage of bromphenol blue from the ventricle to the adjacent nervous tissue was very much less in the dead, than in the living animal and they concluded from this that the process was "active". As Rall has pointed out (personal communication) this is not a justifiable conclusion, since in the dead animal the extracellular space diminishes rapidly (van Harreveld and Ochs 1956) so that the pathway for diffusion from the ventricle through the tissue is much restricted.

IV. The secretory process in the choroid plexuses

A. Morphology

The electron microscope has revealed a number of interesting details of the
structure of the epithelium and blood vessels of the choroid plexuses, although
their significance is not yet clear. DEMPSEY and WISLOCKI (1955) described
the cuboidal cells of the epithelium resting on a basement membrane of electron-
dense material; between this and an adjacent capillary there was connective
tissue consisting of collagenous fibrillae; the capillary endothelium also rested on
a basement membrane. MAXWELL and PEASE (1956) and PEASE (1956) drew
attention to special features of the epithelial cells; at their apical ends, projec-
ting into the ventricle, they resembled the brush-border cells of intestinal epithelium, containing a vast number of processes which were presumably very
watery as they had little electron-density; MAXWELL and PEASE preferred to
call the border "polyploid" rather than "brush", and suggested that these processes might break off as blebs during the secretory process. At their basal ends
the cells showed characteristic infoldings of their plasma membranes that had
also been observed in the kidney tubule cells by SJÖSTRAND and RHODIN (1953)
and the inner layer of epithelial cells of the ciliary body (PEASE 1956), i.e. in
epithelia noted for their continuous transport of water. In the choroidal epithelium the folds were mainly in the lateral parts of the basal portion of the cell,
so that there was considerable interdigitation of the surfaces of adjacent cells.

As indicated above, a thin dense basement membrane underlies the capillary endothelium; pial cells tend to interpose sheets of cytoplasm between epithelium and endothelium; these sheets are not continuous, however, so that they might act as a baffle
rather than as a protoplasmic barrier. Non-myelinated nerves were observed in the
meningeal tissue spaces, obviously associated with arterioles; no endings occurred on
the epithelial cells nor yet intracellularly.

WISLOCKI and LADMAN (1958) have drawn attention to changes in the epithelial cells
that may possibly be associated with phases of secretory activity. The endothelial cells
of the capillaries often contained pores or attenuated regions, as described by MAXWELL
and PEASE (1956). WISLOCKI and LADMAN described cilia protruding from the apical
surfaces of the epithelial cells of all species examined (rat, opossum, dog, monkey).

In tissue-culture preparations of choroid plexuses from rat, rabbit and
chick embryos, CAMERON (1953) described the formation of closed cysts; these
cysts were able to accumulate phenol red and orange G when these were added
to the medium, thus indicating the power of these cells to transport these dye-
stuffs actively. When deprived of oxygen the dye-stuffs diffused back into the
medium. Later LUMSDEN (1958) continued this work, showing that the cysts
swelled, presumably as a result of secretion of fluid into them by their lining
epithelial cells. A difference of pH between inside and outside of the cyst was
maintained.

B. Metabolic aspects of secretion

The oxygen consumption of the plexus is high, when it is considered how
much of it is connective tissue, and is comparable with that of liver and

kidney. Enzymological studies indicate a high succinoxidase activity in the epithelial cells, as well as carbonic anhydrase activity (Fisher and Copenhaver 1959) whilst the blood vessels of the plexus showed a high alkaline phosphatase activity (de Sibrik and O'Doherty 1960).

Metabolic poisons may be expected to reduce the rate of formation of cerebrospinal fluid; the great difficulty in this sort of study, however, is to poison the choroid plexuses without at the same time killing the animal. The carbonic anhydrase inhibitor, Diamox, is valuable in this respect since experimental animals and man can tolerate blood concentrations that apparently inhibit the activity of this enzyme completely in a variety of tissues including the choroid plexuses, the gastric epithelium, ciliary epithelium, and so on. Empirically it has been found to inhibit gastric secretion, presumably because a rate-determining step in this process is the conversion of metabolic CO_2 to carbonic acid, which then reacts with OH-ions liberated in the production of acid. Tschirgi, Frost and Taylor (1954) and Kister (1956) showed that the cerebrospinal fluid pressure was reduced by Diamox, an effect that was presumably due to a slowing of formation of the cerebrospinal fluid since the rate at which the fluid dripped from a cannula in the subarachnoid space fell considerably (to 30% of normal) on treatment of the animal with the drug. There is good reason to believe that rate of turnover of ^{24}Na in the cerebrospinal fluid is a measure of the rate of renewal of the cerebrospinal fluid as a whole (Davson 1956); if this is true we may compute a slowing of formation of fluid by about 50% by treatment of a rabbit or dog with Diamox since the rate of turnover of ^{24}Na in the fluid falls by about this amount (Davson and Luck 1957), Fishman (1959). Again, the rate of appearance of iodinated serum albumin in the plasma, after injection into the subarachnoid space, will be determined by the rate of flow of cerebrospinal fluid, i.e. by its rate of production; hence Diamox should reduce the rate of appearance of the labelled protein in the plasma. In fact, van Wart, Dupont and Kraintz (1961) found that the rate was reduced to one half normal and they concluded, correctly, that the phenomenon was the result of a decreased flow of fluid rather than to the poisoning of a hypothetical absorptive mechanism. According to Davson and Luck, Diamox causes the cerebrospinal fluid to become more like a dialysate of plasma, in the sense that the chloride distribution-ratio, R_{csf}, falls from a normal value of 1.17 to 1.13, and this has been confirmed by Maren and Fischer (1959), Maren and Robinson (1960) and Hogben, Wistrand and Maren (1960) in both mammals and fishes.

The effects on the cerebrospinal fluid pressure are rather confusing since there is, at first, a considerable rise in pressure which is apparently due to a primary dilatation of the cerebral vessels consequent on the accumulation of CO_2 in the blood (Coppen and Russell 1957; Knopp, Atkinson and Ward 1957; Atkinson and Ward 1958), but the value in the treatment of hydrocephalus seems well established (BIRZIS, CARTER and MAREN 1958; FISHER and COPENHAVER 1959; ELVIDGE, BRANCH and THOMPSON 1957). TSCHIRGI et al. (1954) and TSCHIRGI (1960) deduced, from their observation of the reduced rate of flow of cerebrospinal fluid resulting from Diamox treatment, that the conversion of metabolic CO_2 of the nervous tissue into carbonic acid was a prime step in the secretion of cerebrospinal fluid, and from this they were led to argue that the nervous tissue itself contributed materially to the actual formation of the fluid which was therefore not exclusively, or even chiefly, the product of the choroid plexuses. If the nervous tissue is indeed secreting an extracellular fluid which is transported into the ventricles and subarachnoid spaces, then the turnover of ^{24}Na in the nervous tissue must also be affected by Diamox. In fact, as DAVSON and LUCK (1957) showed, Diamox has no influence on this. It could be argued that it failed to have an effect because of its slow diffusion through the blood-brain barrier (ROTH, SCHOOLAR and BARLOW 1959) but by using the lipid-soluble neptazane this could be ruled out as an explanation since this inhibitor likewise had no effect on turnover of ^{24}Na in nervous tissue whilst slowing that in the cerebrospinal fluid[1].

If the turnover of ^{24}Na in the fluid is a fair measure of the rate of secretion — and the results with Diamox tend to confirm this view — then the effects of other drugs on the rate of production of cerebrospinal fluid may be studied similarly. Actually FISHMAN (1959) has studied various substances, including cortisone, chlorothiazide and mercurial diuretics. Of the substances studied, only vasopressin caused an accelerated turnover. A preliminary drainage of 7 ml of fluid from the dog caused about a 50% increase in rate of turnover, suggesting that production of fluid is, indeed, accelerated by this procedure.

BERING (1959) has tried to correlate rate of flow of cerebrospinal fluid, drawn from the subarachnoid space under a negative pressure of —50 mm Hg, with the cerebral oxygen consumption. In fact there was a correlation between flow-rate per unit weight of brain and oxygen consumption per unit weight of brain; a second correlation was between flow-rate per unit weight of plexus and cerebral blood flow. From this he concluded that the fluid was formed by two separate mechanisms.

C. Cerebrospinal fluid potentials

Active transport processes are often associated with a gradient of electrical potential across the secretory layer, as for example, the frog skin or gastric mucosa. TSCHIRGI and TAYLOR (1958) have recorded a potential across an electrode placed in the cerebral cortex and another in the jugular vein, and shown that this varies with the pH of the plasma, but the relationship to secretory processes in ventricles or parenchyma is questionable.

---

[1] We may note that BERING (1958) states that he was unable to detect any change in the rate of flow of cerebrospinal fluid from a cannula in the cisterna magna of the dog on treatment with Diamox. Moreover, plexectomized animals (lateral ventricles) also showed no change in rate of production as measured in this rather unphysiological way. However, the evidence so far cited is overwhelmingly in favour of a diminished rate of secretion of cerebrospinal fluid as a result of treating the animal with Diamox, and this has been confirmed recently in the present author's laboratory (POLLAY and DAVSON) using a ventriculo-cisternal perfusion technique with the rabbit.

In the dogfish HOGBEN, WISTRAND and MAREN (1960) have recorded a potential of about 8 mV between electrodes in ventricle and extracellular fluid, the ventricle being negative; they concluded from this that the high value of R_{csf} for the chloride ion must be the result of transport against an electrochemical gradient. The potential was unaffected by Diamox, however.

V. The mechanism of drainage of cerebrospinal fluid

The arachnoid villi have been accepted by most recent workers as the site of absorption of cerebrospinal fluid into the blood; a recent paper by WELCH and FRIEDMAN (1960) has added powerful support to this view, although modifying WEED's original concept as to the mechanism of absorption. According to WEED, the villus was completely capped by membrane material, so that drainage into the blood consisted essentially in a passage through a membrane, the rate of passage being determined by the difference of hydrostatic pressure plus the difference of osmotic pressure between the two fluids. That the colloid osmotic pressure of the plasma proteins would be an effective force pulling

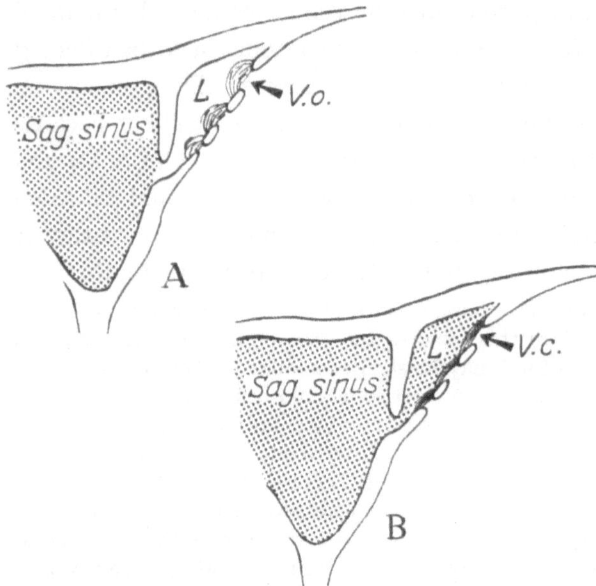

Fig. 8. Diagrammatic illustration of the situation and function of the arachnoid villi. When the meningeal pressure is sufficiently greater than that in the sinus the villus meshwork is open as in (A). When the meningeal pressure is below that in the sinus, reflux of blood is prevented by collapse of the villus (B). (From WELCH and FRIEDMAN 1960)

the cerebrospinal fluid out of the subarachnoid space was questioned by DAVSON (1956), who pointed out that this would require that the membranous covering of the villus be impermeable to proteins, and thus the return of proteins to the blood from the cerebrospinal fluid would be prevented. This difficulty has been resolved by WELCH and FRIEDMAN, who showed, by means of histological preparations of the villi in monkeys, that these could be considered as aggregates of tubes opening directly into the lacunae laterales of the dural sinus provided that the pressure in the subarachnoid space was high enough; when the pressure was too low, the tubes collapsed, so that the system of villi operated as a valvular mechanism for regulating flow from subarachnoid space to dural sinus, as illustrated by Fig. 8. By taking a piece of dura containing numerous villi, and using it as a membrane separating two tubes filled with fluid, the pressure-flow relationships through these valve-like villi could be studied experimentally; the critical pressure difference, below which flow in the direction of the subarachnoid space to dural sinus ceased, was of the order of 25—50 mm saline.

The histology of the arachnoid villi and the pia-arachnoid in the para-sagittal region has been described in some detail by TURNER (1958, 1961); the striking feature of his results is the dense nature of the tissue in the sub-arachnoid space which would cast some doubt on the presence of much cerebrospinal fluid in this particular region.

VI. The cerebrospinal fluid pressure

Very little has been done on this aspect. The effects of hypertonic solutions of urea have been alluded to earlier (p. 40). BEDFORD [1958 (a), (b), 1960] has studied the effects of tilt on the pressures recorded from cisterna magna and lumbar theca. On tilting, the rise or fall in pressure recorded from the cisterna magna is not sustained but rather compensated or, indeed, over-compensated, possibly as a result of altered rate of absorption of cerebrospinal fluid but more probably as a result of adjustments in the venous pressures in cranial and spinal cavities as suggested earlier (DAVSON 1956) when WEED'S studies were discussed. As BEDFORD points out, this compensatory mechanism ensures that the cerebrospinal fluid pressure shall not vary greatly in spite of large alterations in posture. The most striking studies are those on the high cerebrospinal fluid pressure caused by hypovitaminosis-A; the early work has been summarized by WOOLLAM and MILLEN (1956) and MILLEN and WOOLLAM (1958) who consider that the avitaminosis increases the rate of production of fluid, but the raised pressure and hydrocephalus could equally well, and more probably, result from an impairment of the drainage mechanism, and certainly histological studies of the arachnoid villi under these conditions would be interesting.

BERING and SALIBI (1959) have shown that raised cephalic venous pressure may cause hydrocephalus. The rise in cerebrospinal fluid pressure caused by 2 min apnoea may be nearly 100% (SMALL, WEITZNER and NAHAS 1958) and is primarily due to cerebral vasodilatation. LOCKSLEY (1958) has varied the fluid pressure experimentally by connect-ing the cisterna magna to a reservoir of saline. This was accompanied by a flow of saline into the cisterna, and the rate of flow was measured. At a reservoir pressure of 1500 mm saline there was an influx of 300 ml/hr, and this was tolerated for as long as 9 min, the animal apparently recovering completely. The Cushing-reaction occurred only at fluid pressures above 1000 mm saline.

The effects of drugs on the brain-tissue pressure, i.e. the pressure recorded by inserting a needle into the parenchyma, have been described by MALAMANI and CAMPAGNARI (1956).

VII. Chemistry of the cerebrospinal fluid

A. Proteins

From a physiological point of view, the protein composition is of perhaps rather a limited interest, but from the clinician's viewpoint the more detailed our knowledge of the normal components of the proteins, and the variations of these in disease, the better for diagnosis. In general, it is well established that the normal proteins are, with the possible exception of the τ-fraction of the β-globulins, not peculiar to the cerebrospinal fluid but may also be found in the plasma (see, for example, SCHEIFFARTH et al. 1958); abnormal proportions of the different components could arise by the passage of material from the nervous tissue into the cerebrospinal fluid, for example from a tumour. The extent to which the proteins are derived from the choroid plexuses, and from

Table 2. *Protein fractions in normal cerebrospinal fluid expressed as % of total. The τ-fraction has been included with the β-globulins*

		Pre-albumin	Albumin	Globulins			
				α_1	α_2	β	γ
1952	ESSER	1.2	56.1	4.7	7.5	24.1	6.4
1953	BAUER	4.2	59.4	13.4		13.4	9.4
1953	OLDERHAUSEN, FRIES, ALY	4.3	51.3	5.8	8.4	23.9	6.3
1954	KUTZIM, SCHEID, VONKENNEL	4—5	52.0	4.5	8.0	22.0	8.9
1954	SCHÖNENBERG		58.7	7.7	9.9	12.9	10.9
1955	KNAPP	4.3	46.0	7.4	10.5	22.2	9.6
1956	DELANK	1.9	53.4	6.4	8.4	20.0	9.9
1956	SARTESCHI, FABIANI	2.4	52.8	6.8	9.1	19.7	9.2
1956	SCHMIDT, MATIAR	6.4	48.6	7.5	9.4	19.9	8.2
1957	MUMENTHALER, MÄRKI	4.4	49.2	6.5	8.1	26.2	5.6
1957	RISIO, INESI, TONINI	7.3	45.2	13.4		20.3	13.8
1957	ROSSI	1.2	55.3	7.0	5.5	21.1	9.5
1958	BÉLANGER, MARTIN	3.4	57.7	4.9	7.5	18.1	10.1
1958	HANZAL, VYMAZAL	2.9	59.8	4.5	5.6	17.4	8.8
1960	SPINA-FRANÇA	2.2	51.6	5.0	8.7	21.6	10.9

the nervous tissue, is one of the problems towards which research is being directed.

Refinements of electrophoretic technique (see, e.g. KARCHER, VAN SANDE and LOWENTHAL 1959) have led to great improvements in the accuracy with which the individual components of the proteins in the cerebrospinal fluid can be resolved. A valuable recent paper on the electrophoretic analysis of cerebrospinal fluid is that of SPINA-FRANÇA (1960), who has not only reported on his own studies but has also compared his results with earlier ones and his Table is reproduced above (Table 2).

SPINA-FRANÇA was at pains to compare his values with those in the serum of the same patients; the results for seven subjects, expressed as the ratios of the percentages of the given protein in the total, are shown in Table 3.

Here the "pre-albumin" of cerebrospinal fluid has been lumped with the albumin fraction and the τ-fraction with the β-globulins. The results suggest that the cerebrospinal fluid contains an outstandingly high proportion of β-globulin and a low γ-globulin proportion by comparison with serum.

One of the most exhaustive studies on normal ventricular, cisternal and lumbar fluids is that of HILL, McKENZIE, GOLDSTEIN, McGUCKIN and SVIEN (1958). Some figures on the normal components are given in Table 4. In general, when ventricular, cisternal and lumbar fluids were compared, the absolute concentrations of protein increased in this order (see also, HUNTER and SMITH 1960); the relative proportions of the different components were the same in the fluids from the three sources only when the concentration in the lumbar fluid was in the low range of normal[1] (10—34 mg/100 ml); when the lumbar fluid had a high, but still within normal limits, concentration, the proportions changed, indicating that the high concentration in the lumbar fluid was not just a result of concentration of the fluid by absorption of water, as had been suggested. In general, it appeared that the high concentration was due to the appearance of more albumin and (to a less extent) α_1- and γ-globulins. Thus we may envisage the normal secretion of a primary cerebro-spinal fluid containing constituent proteins in characteristic proportions, which are not necessarily the same as those in the original plasma; on its passage through the ventricles and subarachnoid spaces additional proteins diffuse

Table 3. *Ratio of the concentrations of proteins in cerebrospinal fluid and serum in human subjects* (SPINA-FRANÇA 1960)

Protein	% in csf/ % in Serum
Albumin	1.0
α-Globulin	0.9
β-Globulin	1.6
γ-Globulin	0.6

Table 4. *Proteins in ventricular, cisternal and subarachnoid fluids. Total proteins are given as mg/100 ml, whilst the fractions are given as percentages of the total* (HILL et al. 1958)

Site	Total	Pre-albumin	Albumin	Globulins			
				α_1	α_2	β	γ
Ventricle	17.1 ± 9.9	6.3 ± 1.8	46.6 ± 6.5	8.1 ± 1.7	7.9 ± 2.8	19.1 ± 2.0	10.3 ± 3.7
Cistern	18.3 ± 4.3	4.6 ± 1.6	44.6 ± 7.3	6.7 ± 1.0	9.5 ± 3.7	21.3 ± 4.5	13.4 ± 4.0
Lumbar	21.0 ± 7.3	5.5 ± 1.1	45.0 ± 6.0	6.8 ± 2.6	9.6 ± 2.3	21.6 ± 4.7	11.1 ± 2.7
	39.5 ± 4.8	3.9 ± 0.8	46.4 ± 3.6	6.3 ± 2.4	10.5 ± 2.6	18.8 ± 3.7	12.6 ± 3.7
	59.1 ± 3.8	3.3 ± 1.2	53.6 ± 5.6	5.9 ± 1.6	8.6 ± 1.2	17.6 ± 1.2	10.0 ± 2.5

from the tissue into the fluid, the amounts of the different types depending perhaps on the ease with which these escape from the blood into the tissue, or from the blood in the pia into the subarachnoid space, and also on the breakdown of material in the tissue.

This view of a gradual enrichment of the fluid during its passage from choroid plexuses to arachnoid villi was substantiated by the study of fluid below a block (HILL, GOLDSTEIN, McKENZIE, McGUCKIN and SVIEN 1959).

---

[1] The concentration of pre-albumin is greatest in the ventricles, less in the cisterna magna and least in the lumbar region (STEGER 1953); thus it is only by lumping the pre-albumin fraction with the albumin that we can say that the proportions of the components are the same in ventricular and subarachnoid fluids.

In agreement with earlier work it was found that the protein content rises below the block; the analysis showed that this rise was not due to a concentrating process, i.e. removal of water, since the percentage of pre-albumin actually fell from 4.6 to 1.0%, whilst the proportion of albumin rose from 49.5 to 56.8% in partial, and 53.5% in complete block. To a less extent α_1-globulin and γ-globulin increased their proportions. The authors suggest that the ventricles are the source of pre-albumin and this would account for their high proportion in ventricular fluid; it would presumably diffuse slowly into the nervous tissue across the linings of the subarachnoid space. Meanwhile small amounts of plasma proteins would diffuse from the nervous tissue into the subarachnoid fluid, and albumin, being the smallest of the protein molecules, would naturally be favoured. Under abnormal conditions, when the fluid was loculated below a block, the process of diffusion would continue and, besides increasing the absolute concentrations of the plasma proteins, the relative differences, only faintly indicated in the ventricular-lumbar differences in the normal, would be accentuated.

In newborn normal infants the concentration of proteins is high; according to PILLIERO and LENDING (1959) the average for 35 full-term infants, from $2^1/_2$ to 114 hours old, was 70 ± 6 mg/100 ml comparing with the normal adult value of 15—30 mg/100 ml. The proportions of the different fractions were: pre-albumin, 2.5; albumin, 47.4; α_1-globulin, 6.8; α_2-globulin, 8.8; β-globulin, 14.5; γ-globulin, 20.0%. In older children LANZA and LOMBARDO (1959) described values of 0.25 to 0.35% for total proteins, i.e. 250—350 mg/ 100 ml, but their figures were probably in error by a factor of 10.

Pre-albumin. This rapidly migrating component of the cerebrospinal fluid proteins is not normally observed in electrophoretic studies of plasma proteins; it is present, however, being some 0.5% of the total (SCHULTZE, SCHÖNEN-BERGER and SCHWICK 1956); from light-scattering data its molecular weight is 61000, and in its carbohydrate and amino acids it is characteristically different from serum albumin. According to BOOIJ (1958) it is always present in brain extracts and he concludes that the brain is the source of that found in the cerebrospinal fluid, but if this were true one would not necessarily expect to find the highest proportion in the ventricular fluid.

Because of the difficulty in obtaining cerebrospinal fluid from different parts of the cerebrospinal system, it is interesting to determine whether the examination of successive samples of fluid, withdrawn from the lumbar region during ventriculography, can contribute to our knowledge. According to GEINERT and MATIAR (1959), the later phases of withdrawal, i.e. after the removal of some 70 ml, are accompanied by a large increase in the percentage of β-globulin, and a fall in that of pre-albumin. They concluded that the rise in the percentage of β-globulin was due to the arrival of fluid from the cranial subarachnoid space, where the concentration of protein is considered to be highest. Thus, the rise in percentage of β-globulin in the cerebrospinal fluid that occurs in brain atrophy may be due to an increased volume of cranial subarachnoid fluid rather than to any unusual influx of this globulin into the cerebrospinal fluid.

Labelled protein studies. FISHMAN, RANSOHOFF and OSSERMAN (1958) injected ^{131}I-labelled serum albumin intravenously into human infants and measured the rate of appearance of radioactivity in the cerebrospinal fluid; they found that the order of rates of appearance was lumbar > cisterna > ventricle. This strongly suggests that protein normally penetrates into the cerebrospinal fluid after it has left the ventricles, presumably by way of the nervous tissue. A more elaborate study of blood-cerebrospinal fluid relationships is that of FRICK and SCHEID-SEYDEL [1958 (a), (b), 1960], who injected labelled albumin and γ-globulin intravenously into human subjects and determined the specific activities of the proteins in serum and cerebrospinal fluid at different intervals. They found that some 3 days were required for the establishment of a dynamic equilibrium between serum and cerebrospinal fluid, in the sense that the specific activities of the albumin in the two fluids reached a steady relationship at the end of this period. Because of the breakdown of serum protein the specific activity in the serum decreased continuously, but that in the serum, after the establishment of the dynamic equilibrium, was always some 20% higher, showing that breakdown of albumin did not take place in the cerebrospinal system, which behaved similarly in this respect to ascites. With γ-globulin, some 100 hours were required for the establishment of the dynamic equilibrium. With both albumin and γ-globulin it was concluded, because of the higher specific activity in the cerebrospinal fluid, that the proteins were entirely derived from the blood; in cases where the γ-globulins were pathologically elevated, however, the specific activity could be only 10% of that in the serum, suggesting that the abnormal level was due to influx from the nervous tissue rather than the blood.

When β-globulin was labelled (FRICK and SCHEID-SEYDEL 1960) it was found that the specific activity in the cerebrospinal fluid, when dynamic equilibrium was established, was *greater* than in serum, indicating the presence in the cerebrospinal fluid of a β-globulin of cerebral origin. This accords with the finding that the cerebrospinal fluid normally contains a higher percentage of β-globulin than that in the serum; that it contains a β_2- or τ-fraction not found in the serum; and also with the finding that the concentration of β-globulin is higher in the cranial subarachnoid fluid than in the lumbar fluid; and finally with the fact that when the blood-brain barrier is broken down, as in meningitis, the percentage of β-globulin in the cerebrospinal fluid falls.

The penetration of iodinated bovine albumin from blood to brain-tissue has been studied in the rabbit, cat and dog by LEE and OLSZEWSKI (1960); it seemed to enter diffusely into the subpial tissue as well as along the blood vessels, reaching a peak-concentration some 3 hr after the injection.

Absorption from cerebrospinal fluid. The absorption of labelled serum albumin from the subarachnoid space has been described by several workers, and the results are consistent with the view that most of the absorption is by way of the main drainage route in the subarachnoid space, i.e. the arachnoid villi (SWEET et al. 1954). Thus ISHIBASHI (1959) injected labelled albumin into the cranial subarachnoid space and found it appeared more rapidly in the blood when injected here than if it had been injected in the ventricles or lumbar subarachnoid space. When injected into the ventricle of a dog whose aqueduct had been blocked, the appearance in the blood was very slow. By following activity with an external counter, the gradual migration of labelled protein from ventricle to cisterna magna and lumbar region was followed.

VAN WART, DUPONT and KRAINTZ (1960) followed the rise in the radio-activity of the plasma after injecting labelled albumin into the cerebrospinal fluid; it required some 22—24 hr for this level to reach a steady state and they computed a half-life for absorption of some 6 hr, but it seems unlikely that this represents the half-life for the turnover of the cerebrospinal fluid as a whole since many other studies suggest a rate of renewal of the order of 0.5 % per min; this corresponds to a turnover constant k, of 0.005 min^{-1}; since $t_{\frac{1}{2}}$, the half-life, is related to the turnover constant by:

$$t_{\frac{1}{2}} = \frac{\log 2}{k}$$

the half-life of the cerebrospinal fluid is about 0.7/0.005, i.e. 140 min[1].

That absorption is not entirely from the arachnoid villi is indicated by BOWSHER's (1957) radioautographic study of cats injected with ^{131}I-labelled albumin; blackening was observed on the cerebral and spinal pia and in the underlying tissue suggesting (to BOWSHER) direct absorption into the blood vessels; the difficulty in accepting a direct absorption into the blood, of course, is the circumstance that the gradient of concentration is in the wrong direction, so that the demonstration that labelled protein may pass from cerebrospinal fluid to blood vessel rather proves that the net flux of protein must be in the reverse direction, i.e. that the blood vessels of pia and paren-chyma add to the protein in the fluid.

Glycoproteins. By glycoprotein (WINZLER 1955) is meant any combination between protein and carbohydrate. They may be classed as *mucoproteins*, where the linkage is relatively weak, as for example, the combination between

---

[1] Only if the albumin leaving the subarachnoid space entered a single compartment, namely the plasma, would the rate of rise in concentration of the plasma be a direct measure of the rate of escape from the cerebrospinal fluid. The subsequent passage into the extracellular space, and the urinary excretion, obviously complicate the issue, so that analysis of the cerebrospinal fluid is preferable as in the experiments of SWEET et al. (1954). We may note that CHOU and FRENCH (1955) found that the blood level reached its peak in about 40 hr after injection into the lumbar theca. Using an external counter, these authors examined the passage of material up and down the cord after injection at L$_4$-L$_5$. After one hour, activity was measurable in the cisterna magna, but it required some 20 hr for equilibrium to be established, in the sense that the relative activities in the different regions remained constant.

chondroitin sulphate and protein found in cartilage, and *mucoids*, where the linkage is stronger and probably covalent. In the serum there are numerous *seromucoids*, some of which have the same electrophoretic mobilities as the ordinary proteins, for example, *orosomucoid* has the mobility of γ-globulin. Sialic acid, or neuraminic acid, is a substance found in combination in these seromucoids. In general, an increase in the percentage of glycoproteins occurs in many diseases, e.g. tuberculosis, neoplasms, and so on, and this may well be due to the breakdown of tissue or else be a by-product of the rapid synthesis of new tissue as in a neoplastic disease. The presence of glycoprotein in normal

Table 5. *Proteins and glycoproteins of cerebrospinal fluid (csf) and serum of man* (HILL et al. 1958)

	Fluid	Total	Pre-albumin	Albumin	Globulins			
					α_1	α_2	β	γ
Protein	Serum	6.84	—	56	4.9	9.5	12.1	17.5
	csf	31.3	4.6	49.5	6.7	8.3	18.5	11.2
Glycoprotein	Serum	93.0	—	14.3	18.4	29.8	20.5	16.7
	csf	1.0	—	25.0	21.5	18.8	33.5	—

The total proteins of serum are given as g/100 ml, whilst the glycoproteins for both fluids and total proteins for cerebrospinal fluid are given as mg/100 ml. The amounts of the various fractions are given as percentages of the total.

and pathological cerebrospinal fluid has been described by several workers, and their findings have been tabulated by SILVERSTEIN and GREENSPAN (1959). HILL et al. (1958) extended their elaborate study of the protein content of human cerebrospinal fluid to an analysis of the relative proportions of glycoproteins; some of their results are reproduced in Table 5 where serum values are included for comparison. According to these figures, the proportions of glycoproteins are not the same as in the serum. Other studies on the glycoproteins are those of BAUDOUIN, LEWIN and HILLION (1954); ROBOZ, MURPHY, HESS and FOSTER (1955), ROBOZ, APOSTOL, LUFT and HESS (1957), and PADMAVATI, BAKSHI, PANTULU and SINGH (1958).

Neuraminic acid. This substance, which may be identical with sialic acid, is part of the molecular structure of the water-soluble glycolipids known as *gangliosides*, and is associated with the α_1- and α_2-globulin groups of glycoproteins. Neuraminic acid found in the cerebrospinal fluid is partly bound to protein whilst the greater portion is dialysable, in contrast with serum where the whole is bound to protein (UZMAN and RUMLEY 1956). When different cerebrospinal fluids are examined, the amount of bound neuraminic acid increases passively with the protein content. The dialysable neuraminic acid is highest in the ventricular fluid and lowest in the lumbar; it is low in the fluid loculated below a spinal block and high in the ventricular fluid in non-communicating hydrocephalus due to obstruction at the level of the IVth ventricle (UZMAN, BERING and MORRIS 1959). Its association with the ventricles suggests that the dialysable neuraminic acid is either secreted by the choroid plexuses or else is a product of cerebral metabolism. According to BOGOCH (1958), the concentration is lower than normal in the cerebrospinal fluid of schizophrenics.

Enzymes. The presence, and degree of activity, of several enzymes, in particular glutamic oxalacetic-transaminase and lactic dehydrogenase, have come in for a lot of study recently because changes in these parameters seem to reflect certain pathological conditions. It is beyond the scope of this review to describe these studies, and at the risk of being invidious only two papers, which summarize the literature on the subject, will be quoted, namely those of HAIN and NUTTER (1960) and GREEN and OLDEWURTEL (1960).

B. Amino acids

The rates of exchange of amino acids between plasma and brain and cerebrospinal fluid are relatively rapid (LAJTHA 1959; LAJTHA and MELA 1961), so we may expect the cerebrospinal fluid to reflect the composition

Table 6. *Mean concentrations of aminoacids (mg/100 ml) in cerebrospinal fluid (csf) of 25 human subjects* (MÜTING and SHIVARAM 1960). *Serum values have been taken from* STEIN and MOORE (1959)

	Csf	Serum
Aspartic acid	0.13	0.03
Glutamic acid	0.27	0.70
Lysine	0.36	2.72
Arginine	0.20	1.51
Histidine	0.26	1.15
Tyrosine	0.18	1.03
Tryptophan	0.20	1.11
Phenylalanine	0.19	—
Hydroxyproline	0.00	—
Proline	0.27	0.84
Cystine	0.24	1.18
Methionine	0.13	0.38
Leucine	0.23	1.69
Isoleucine	0.12	0.89
Valine	0.36	2.88
Glycine	0.39	1.54
Alanine	0.34	3.41
Serine	0.28	1.12
Threonine	0.18	1.39
Taurine	0.32	0.55
Glutamine	0.45	8.30
α-Aminobutyric acid	traces	0.30
γ-Aminobutyric acid	traces	

of the plasma in this respect, but the extent to which the individual acids are involved in the metabolism of the adjoining parenchyma may well modify the "spectrum" of acids in this fluid (LAJTHA, BERL and WAELSCH 1959). Recent exhaustive quantitative studies are those of RAO and GOVINDASWAMY (1958), LOGOTHETIS (1958) and MÜTING and SHIVARAM (1960), whilst a more qualitative study is that of KNAUFF, MIATKOWSKY and ZICKGRAF (1959). The findings of MÜTING and SHIVARAM are given in Table 6; in agreement with others the concentration of glutamine is by far the highest; next come glycine, alanine, valine and lysine. Unfortunately the concentrations in serum of the same subjects were not determined by MÜTING and SHIVARAM, so that no accurate comparison is possible. The figures for human serum given by STEIN and MOORE (1954) are, however, presented so that the orders of magnitude of the concentrations in the two fluids may be appreciated. In general, the concentrations in the cerebrospinal fluid are much lower than in serum; thus the total amino-acid *nitrogen* of cerebrospinal fluid is 0.75 mg/100 ml comparing with a value of some 5 mg/100 ml for serum.

C. Glucose

The low concentration of glucose in cerebrospinal fluid has been discussed earlier in another context. The fall in this that takes place in bacterial meningitis is well known but its cause is not entirely clear. Incubation of cerebrospinal fluid with leucocytes certainly causes a decrease in its glucose concentration, but incubation with bacteria usually does not, yet pathologically it is found that it is the *bacterial* meningitis that is accompanied by a fall in cerebrospinal fluid sugar whilst an aseptic meningitis is not (see, for example, BALTCH and OSBORNE 1957). The contradiction is apparently resolved by recent studies of PETERSDORF, SWAMER and GARCIA [1960 (a), (b)], who have shown that there is a synergistic action between leucocytes and bacteria, so that leucocytes will cause the sugar concentration to fall only when they have bacteria to phagocytose. Thus dogs were given an aseptic meningitis by injection of NaCl into the subarachnoid space; if, now, they were injected with pneumococci, there was a profound fall in glucose concentration, from 80 to 20 mg/100 ml in 7 hr. By contrast, the cerebrospinal fluid from dogs with aseptic meningitis without the pneumococcal injection, or the cerebrospinal fluid from normal dogs with the pneumococcal injection, retained a nearly normal glucose content[1].

D. Inositol

The concentration of this substance in cerebrospinal fluid is very much higher than in plasma (BAUMGARTEN 1958; NIXON 1959) and this accumulation, which presumably is the result of active transport mechanisms by the choroid plexuses, occurs in the foetal sheep of 94 days and older. By contrast, the aqueous humour only shows a similar accumulation in the adult, thus emphasizing the much more rapid development of maturity in the cerebrospinal system.

E. Ions

It has been well established that the concentration of Mg is normally higher in the cerebrospinal fluid than in a dialysate of plasma; thus the value of R_{csf} for this ion is in the range 1.2 to 1.3, whilst, because of the binding to plasma proteins, the value of the dialysis ratio, R_{Dial}, is only 0.785; hence there is

---

[1] Insulin, injected intracisternally in dogs, lowers the concentration of sugar in both cerebrospinal fluid and plasma, that in the cerebrospinal fluid to the greater extent (CHOWERS, LAVY and HALPERN 1961). Using ^{131}I-labelled insulin the authors were unable to measure any influx into the blood, so it was assumed that the action of the hormone was exerted directly on some central nervous structure.

a considerable excess of Mg in the cerebrospinal fluid over that which would be expected were this fluid in thermodynamic equilibrium with the plasma. The values of R_{csf} and R_{Dial} for Ca are related in the opposite way, R_{csf} being 0.4 and R_{Dial} being 0.6, so that normally Ca is in deficit in the cerebrospinal fluid. The very low value of R_{Dial} is due to the well known high degree of binding of Ca to the plasma proteins. Potassium is similar to Ca in having a low value of R_{csf}, namely in the region of 0.6 which compares with a high value of R_{Dial}, namely 0.96, because there is no significant binding of K to plasma proteins. The chloride ion has a higher concentration in cerebrospinal fluid than would be expected of a dialysate ($R_{csf} = 1.15$ about, compared with 1.03 for R_{Dial}); the same is true for Na ($R_{csf} = 1.01—1.04$ compared with 0.95 approx. for R_{Dial}).

There have been several studies on the ionic composition of cerebrospinal fluid in the period under review. KOBAYASHI and KODAMA (1955) have given the following values for R_{csf}[1]:

Ca	Mg	K	Na	Cl
0.61	1.12	0.61	1.03	1.10

the concentrations of the respective ions in the cerebrospinal fluid being correlated with the concentrations in the plasma. KARCHER, LOWENTHAL and VAN SANDE (1957) studied the Ca, K and Na in normal and meningitic fluids but did not compare them with the plasma values; according to these workers, a rise in fluid Ca is usually accompanied by a rise in K, whilst the concentrations of Na and K varied independently. HUNTER and SMITH (1960) have provided a valuable contribution to our knowledge on Ca and Mg distributions; their mean results for normal fluids are as follows:

	Csf	Calcium serum	R_{csf}	Csf	Magnesium serum	R_{csf}
Mean	4.56	9.41	0.486	2.71	1.96	1.386
Range	3.9—5.1	8.2—10.6	0.38—0.56	2.4—2.9	1.6—2.3	1.15—1.63
S.D.	0.32	0.64	0.039	0.11	0.13	0.10

The level in the cerebrospinal fluid seemed to be independent of that in the serum; in tubercular meningitis with high concentrations of protein in the fluid the concentration of Ca rose whilst that of Mg fell, in accordance with "Cohen's Law of Meningitis".

---

[1] The values have been corrected for the higher percentage of solids in plasma than in cerebrospinal fluid (see DAVSON 1956, p. 62).

When the lumbar, cisternal and ventricular fluids were compared the following figures were obtained:

	Lumbar	Cisternal	Ventricular
Ca	4.65	4.49	4.56
Mg	2.66	2.60	2.72
Protein	59	36	15

It would be hazardous to state that the figures showed any trend on passing from ventricular to lumbar fluid. In this connexion we may note that WEISE (1959) found no significant difference between cisternal and lumbar fluids in respect to Na and K concentrations; the concentration of Ca in the cisternal fluid was considerably less than in lumbar fluid (3.8 compared with 5.3 mg/ 100 ml).

PAUPE (1957) has made an interesting comparison of the total and ionized (diffusible) Ca, the latter being assayed by the action of the fluid on the isolated heart. He found no correlation between the total Ca concentrations in cerebrospinal fluid and serum, but if the *ionized* Ca was assayed there was a very good correlation between the concentrations in cerebrospinal fluid and serum[1].

VIII. Miscellaneous aspects

Some miscellaneous studies, not readily incorporated under the headings employed in this review, may now be briefly summarized. The relative *vascularity* of different parts of the central nervous system was determined by BARLOW, SCHOOLAR and ROTH (1958) using autoradiographic analysis of the tissue after the intravenous injection of labelled serum albumin. The relationship of vascularity to the uptake of intravenously injected ^{32}P-labelled inorganic phosphate was examined by BAKAY (1957). The *water contents* of different parts of the brain and cord have been studied by DOMEK, BARLOW and ROTH (1960); in the newborn kitten this is reasonably uniform and it requires some 12 weeks for the characteristic difference between white and grey matter to be fully established. The *incorporation of* ^{32}P into different layers of cortex has been examined by BAKAY (1959); although metabolic activity must be concerned in this, it would appear that the time course of events is largely determined by the physical factor of diffusion from the cerebrospinal fluid. The effects of metabolic poisons on the uptake of ^{32}P from the subarachnoid space have been described by HERLIN (1958) and ERNSTER and HERLIN (1961). The effects of a variety of insults on the *blood-brain barrier* have been described: LEE (1959), air embolism; BRIERLEY (1956), local cold-injury; BARLOW (1956) and VULPE, HAWKINS and ROZDILSKY (1960), allergic response; SLOBODY et al.

---

[1] HARRIS and SONNENBLICK (1955) and HARRIS and BEAUCHEMIN (1956) have reported on the Ca/Mg ratios in the cerebrospinal fluid of normal subjects and those with psychoses of organic and functional origin. The ratio was remarkably constant being 1.64 $\pm$ 0.03 when the concentrations are expressed in mg/100 ml. The concentrations of Na, K, Cl and H_2O in six separate regions of the brain substance have been measured by APRISON, LUKENBILL and SEGAR (1960). DOS REIS, DOS REIS and BEI (1957) have compared diffusible and non-diffusible Ca in lumbar and cisternal fluids, the differences being hardly significant. WORATZ and ROTZSCH (1960) have analysed the concentrations of Na, K and Ca in human subjects statistically and conclude that Na and K are independent of disease states.

(1957) and Lending, Slobody and Mestern (1961), hypoxia; Clemedson, Hartelius and Holmberg (1958) and Goldberg, Barlow and Roth (1961), hypercapnia; Lourie, Weinstein and O'Leary (1960), hypothermia; Samorajski and Moody (1957), exposure of brain; Lee and Olszewski (1961), electroshocks. The effects of powerful doses of α-particle radiation on the grey and white matter have been described most recently by Klatzo, Miquel, Tobias and Haymaker (1961) who summarize the earlier literature. The blood-brain barrier to various substances has been measured: Samachson et al. (1959), Ca and Sr; Weil-Malherbe, Axelrod and Tomchick (1959) and Draškoci, Feldberg and Haranath (1960), adrenaline; Dingman and Sporn (1959), proline and derivatives; Ford (1959), triiodothyronine; Draškoci et al. (1960), histamine; Rall and Zubrod, sulphanilic acid and PAH in the dogfish. The effect of *altered composition of the cerebrospinal fluid* on the physiology of the organism has been studied by Loeschke, Koepchen and Gertz (1958); Loeschke and Koepchen [1958 (a), (b)] and Feldberg and Malcolm (1959). The fate of red blood cells, injected into the cerebrospinal fluid, has been described by Adams and Prawirohardjo (1959). Bakay (1957, 1959) has continued his studies on uptake of ^{32}P by brain. The phenol red excretion test in hydrocephalus has been studied by Laurence (1959).

IX. Summary and perspectives

To summarize briefly, we may state that the general concepts of the formation, circulation and drainage of cerebrospinal fluid, as developed by Weed early in this century, remain essentially unchallenged by the work that has succeeded his pioneering studies. The choroid plexuses form the fluid by processes that are now covered by the term "active transport"; the formation is continuous and drainage takes place through the arachnoid villi which apparently have a valvular connection with the blood in the dural sinuses. Modern work has brought out the kinetic significance of the so-called "barriers" and has put them on a firm experimental basis.

In spite of rapid progress in recent years there are two aspects that require considerably more experimentation; one is the relationships of the various fluid compartments — cerebrospinal fluid, intracellular fluid, extracellular fluid, blood plasma — with each other in the normal condition and in oedema. This requires further patient quantitative study of the exchanges of ions and non-electrolytes between these compartments with a view to determining to what extent metabolic factors, besides simple diffusional factors, are operative in determining the relative volumes of the compartments. Such studies must be supplemented with more work on isolated brain *in vitro*. The other aspect, the nature of the active-transport processes concerned in the elaboration of the fluids, also deserves intensive study since ultimately the information gained will lead to methods of control of secretion.

References

Adams, J. E., and S. Prawirohardjo: Fate of red blood cells injected into cerebrospinal fluid pathways. Neurology (Minneap.) 9, 561—564 (1959).

Aprison, M. H., A. Lukenbill and W. E. Segar: Sodium, potassium, chloride and water content of six discrete parts of the mammalian brain. J. Neurochem. 5, 150—155 (1960).

ATKINSON, J. R., and A. A. WARD: Effect of diamox on intracranial pressure and blood volume. Neurology (Minneap.) **8**, 45—50 (1958).

BAKAY, L.: Embryonic development of the barrier. Arch. Neurol. Psychiat. (Chic.) **70**, 30—39 (1953).

— The blood-brain barrier with special regard to the use of radioactive isotopes. Springfield, Illinois: Thomas 1956.

— Relationship between cerebral vascularity and ^{32}P uptake. Arch. Neurol. Psychiat. (Chic.) **78**, 29—36 (1957).

— Radioactive phosphate concentration in the cerebral cortex. Neurology (Minneap.) **9**, 18—23 (1959).

— Studies in sodium exchange. Neurology (Minneap.) **10**, 564—571 (1960).

— H. T. BALLANTINE and E. BELL: ^{32}P uptake by normal and ultrasonically irradiated brain tissue from cerebrospinal fluid. Arch. Neurol. (Chic.) **1**, 59—67 (1959).

— T. F. HUETER, H. T. BALLANTINE jr. and D. SOSA: Ultrasonically produced changes in the blood-brain barrier. Arch. Neurol. Psychiat. (Chic.) **76**, 457—467 (1956).

BALTCH, A., and W. OSBORNE: Inquiry into causes of lowered spinal fluid sugar content: in vivo and in vitro observations. J. Lab. clin. Med. **49**, 882—889 (1957).

BARLOW, C. F.: A study of abnormal blood-brain permeability in experimental allergic encephalomyelitis. J. Neuropath. exp. Neurol. **15**, 196—207 (1956).

— N. S. DOMEK, M. A. GOLDBERG and L. J. ROTH: Extracellular brain space measured by ^{35}S sulphate. Arch. Neurol. (Chic.) **5**, 102—110 (1961).

— J. C. SCHOOLAR and L. J. ROTH: An autoradiographic demonstration of the relative vascularity of the central nervous system. J. Neuropath. exp. Neurol. **17**, 191—198 (1958).

BAUDOUIN, A., J. LEWIN and P. HILLION: Electrophorèse sur papier des protéines du liquide céphalorachidien. Etudes des lipoprotéines. C.R. Soc. Biol. (Paris) **148**, 1033—1036 (1953).

BAUMGARTEN, F.: Papierchromatographische Untersuchungen über Kohlenhydrate und meso-Inosit im Liquor cerebrospinalis. Dtsch. Z. Nervenheilk. **177**, 552—562 (1958).

BEDFORD, T. H. B.: The effects of tilting from the horizontal to the tail upwards position on the pressure of the cerebrospinal fluid of the dog. J. Physiol. (Lond.) **141**, 3—4P (1958a).

— The effect of tilting from the horizontal to the head upwards position on the pressure of the cerebrospinal fluid in the cisterna magna of the dog. J. Physiol. (Lond.) **145**, 9—10P (1958b).

— The effect of tilting from the horizontal to the head upwards position on the pressure of the cerebrospinal fluid in the lumbar subarachnoid space of the dog. J. Physiol. (Lond.) **151**, 31—32P (1960).

BEHNSEN, G.: Über die Farbstoffspeicherung im Zentralnervensystem der weißen Maus in verschiedenen Alterszuständen. Z. Zellforsch. **4**, 515—572 (1927).

BENNETT, H. S., J. H. LUFT and J. C. HAMPTON: Morphological classification of vertebrate blood capillaries. Amer. J. Physiol. **196**, 381—390 (1959).

BERG, O., and B. KALLÉN: Studies on rat neuroglia cells in tissue culture. J. Neuropath. exp. Neurol. **18**, 458—467 (1959).

BERING, E. A.: Composition of cerebrospinal fluid under varying conditions. Neurology (Minneap.) **8**, Suppl. 129—130 (1958).

— Cerebrospinal fluid production and its relationship to cerebral metabolism and cerebral blood flow. Amer. J. Physiol. **197**, 825—828 (1959).

—, and N. AVMAN: The use of hypertonic urea solutions in hypothermia. J. Neurosurg. **17**, 1073—1082 (1960).

—, and B. SALIBI: Production of hydrocephalus by increased cephalic venous pressure. Arch. Neurol. Psychiat. (Chic.) **81**, 693—698 (1959).

BIRZIS, L., C. H. CARTER and T. H. MAREN: Effect of acetazolamide on c.s.f. pressure and electrolytes in hydrocephalus. Neurology (Minneap.) **8**, 299—302 (1958).

BOGOCH, S.: Cerebrospinal fluid neuraminic acid deficiency in schizophrenia. Arch. Neurol. Psychiat. (Chic.) **80**, 221—227 (1958).

BOOIJ, J.: The origin of the prealbumin in the cerebrospinal fluid. Folia psychiat. neerl. **61**, 84—90 (1958).

BOSTRÖM, H., and E. ODEBLAD: Autoradiographic observations on the uptake of S^{35} labelled sodium sulphate in the nervous system of the adult rat. Acta psychiat. (Kbh.) **28**, 5—8 (1953).

BOURDILLON, R. B., M. FISCHER-WILLIAMS, H. V. SMITH and K. B. TAYLOR: The entry of radiosodium and of bromide into human cerebrospinal fluid. J. Neurol. Neurosurg. Psychiat. **20**, 79—97 (1957).

BOWSHER, D.: Pathways of absorption of protein from the cerebrospinal fluid: an autoradiographic study in the cat. Anat. Rec. **128**, 23—40 (1957).

BRADBURY, M. W. B., and R. V. COXON: The penetration of urea into the central nervous system at high blood levels. J. Physiol. (Lond.) **163**, 423—435 (1962).

BREEMEN, V. L. VAN, and C. D. CLEMENTE: Silver deposition in the central nervous system and the haematoencephalic barrier studied with the electron microscope. J. biophys. biochem. Cytol. **1**, 161—165 (1955).

BRIERLEY, J. B.: The prolonged and distant effects of experimental brain injury on cerebral blood vessels as demonstrated by radioactive tracers. J. Neurol. Psychiat. **19**, 202—208 (1956).

BRODIE, B. B., and C. A. M. HOGBEN: Some physico-chemical factors in drug action. J. Pharm. Pharmacol. **9**, 345—380 (1957).

— H. KURZ and L. S. SCHANKER: The importance of dissociation constant and lipid-solubility in influencing the passage of drugs into the cerebrospinal fluid. J. Pharmacol. exp. Ther. **130**, 20—25 (1960).

BROMAN, T.: Supravital analysis of disorders in the cerebrovascular permeability. Acta psychiat. (Kbh.) **25**, 19—31 (1950).

CAMERON, G.: Secretory activity of the choroid plexus in tissue culture. Anat. Rec. **117**, 115—123 (1953).

CAMMERMEYER, J.: Reappraisal of the perivascular distribution of oligodendrocytes. Amer. J. Anat. **106**, 197—219 (1960a).

— Is the perivascular oligodendrocyte another element controlling the blood supply to neurones? Angiology **11**, 508—517 (1960b).

CANTI, R. G., J. O. W. BLAND and D. S. RUSSELL: Tissue culture of gliomata. Cinematographic demonstration. Ass. Rev. nerv. Dis. Proc. **16**, 1—24 (1937).

CHOU, S. N., and L. A. FRENCH: Systemic absorption and urinary excretion of Risa from subarachnoid space. Neurology (Minneap.) **5**, 555—557 (1955).

CHOWERS, I., S. LAVY and L. HALPERN: Effect of insulin administered intracisternally in dogs on the glucose level of the blood and cerebrospinal fluid. Exp. Neurol. **3**, 197—205 (1961).

CLEMEDSON, C. J., H. HARTELIUS and G. HOLMBERG: The influence of carbon dioxide inhalation on the cerebral vascular permeability to trypan blue ("the blood-brain barrier"). Acta path. microbiol. scand. **42**, 137—149 (1958).

COOPER, R. G., and J. W. ARCHDEACON: Blood and cerebrospinal fluid glucose in the fasting state. Amer. J. Physiol. **198**, 260—262 (1960).

COPPEN, A. J., and G. F. M. RUSSELL: Effect of intravenous acetazolamide on cerebrospinal-fluid pressure. Lancet **1957**II, 926—927.

COULTER, N. A.: Filtration coefficient of the capillaries of the brain. Amer. J. Physiol. **195**, 459—464 (1958).

CRONE, C.: Om diffusionen af nogle organiske non-elektrolyter fra blod til hjernevaev. København: Ejnar Munksgaard 1961.

DAVSON, H.: A comparative study of the aqueous humour and cerebrospinal fluid in the rabbit. J. Physiol. (Lond.) **129**, 111—133 (1955).

— Physiology of the ocular and cerebrospinal fluids. London: Churchill 1956.

DAVSON, H.: Some aspects of the relationship between the cerebrospinal fluid and the central nervous system. In: The cerebrospinal fluid. Ciba Symp., pp. 189—208. London: Churchill 1958.

— C. R. KLEEMAN and E. LEVIN: Blood-brain barrier and extracellular space. J. Physiol. (Lond.) 159, 67—68 P (1961).

— The blood-brain barrier. Proc. Ist Int. Cong. Pharmacol. Stockholm 1962a. (In the press.)

— Quantitative studies of the passage of different substances out of the cerebrospinal fluid. J. Physiol. (Lond.) 161, 126—142 (1962b).

—, and C. P. LUCK: The effect of acetazoleamide on the chemical composition of the aqueous human and cerebrospinal fluid of some mammalian species and on the rate of turnover of ^{24}Na in these fluids. J. Physiol. (Lond.) 137, 279—293 (1957).

—, and E. SPAZIANI: The blood-brain barrier and the extracellular space of brain. J. Physiol. (Lond.) 149, 135—143 (1959).

DEMPSEY, E. W., and G. B. WISLOCKI: An electron microscopic study of the blood-brain barrier in the rat, employing silver nitrate as a vital stain. J. biophys. biochem. Cytol. 1, 245—256 (1955).

DINGMAN, W., and M. B. SPORN: The penetration of proline and proline derivatives into brain. J. Neurochem. 4, 148—153 (1959).

DOBBING, J.: The blood-brain barrier. Physiol. Rev. 41, 130—188 (1961).

DOMEK, N. S., C. F. BARLOW and L. J. ROTH: An ontogenetic study of phenobarbital-C^{14} in cat brain. J. Pharmacol. exp. Ther. 130, 285—293 (1960).

DOMER, F. R.: Transport of ^{42}K from blood to cerebrospinal fluid in cats. J. Physiol. (Lond.) 158, 366—373 (1961).

DOS REIS, I., J. B. DOS REIS e A. BEI: O calcio no liquido céfalo-raqueano. Arch. Neuro-psiquiat. (S. Paulo) 15, 15—26 (1957).

DRAŠKOCI, M., W. FELDBERG, K. FLEISCHHAUER and P. S. R. K. HARANATH: Absorption of histamine into the blood stream on perfusion of the cerebral ventricles, and its uptake by brain tissue. J. Physiol. (Lond.) 150, 50—66 (1960).

—, and P. S. R. K. HARANATH: Passage of circulating adrenaline into perfused cerebral ventricles and subarachnoidal space. J. Physiol. (Lond.) 150, 34—49 (1960).

EDSTRÖM, R. F. S.: An explanation of the blood-brain barrier phenomenon. Acta psychiat. (Kbh.) 33, 403—416 (1958).

—, and H. E. ESSEX: Swelling of damaged brain tissue. Neurology (Minneap.) 5, 490—493 (1955).

ELVIDGE, A. R., C. L. BRANCH and G. B. THOMPSON: Observations in a case of hydrocephalus treated with Diamox. J. Neurosurg. 14, 628—638 (1957).

ERNSTER, L., and L. HERLIN: The fate of intracisternally injected tracer phosphate in the rabbit, normally and as influenced by sulphydryl reagents. J. Neurochem. 6, 267—276 (1961).

FARQUHAR, M. G., and J. F. HARTMANN: Neuroglial structure and relationships as revealed by electron-microscopy. J. Neuropath. exp. Neurol. 16, 18—39 (1957).

FELDBERG, W., and K. FLEISCHHAUER: Penetration of bromophenol blue from the perfused cerebral ventricles into the brain tissue. J. Physiol. (Lond.) 150, 451—462 (1960).

—, and J. L. MALCOLM: Experiments on the site of action of tubocurarine when applied via the cerebral ventricle. J. Physiol. (Lond.) 149, 58—77 (1959).

FERNANDEZ-MORÁN, H.: Electron microscopy of nervous tissue. In: Metabolism of the nervous system, edit. RICHTER, pp. 1—34. London: Pergamon 1957.

FISHER, R. A., and J. H. COPENHAVER: The metabolic activity of the choroid plexus. J. Neurosurg. 16, 167—176 (1959).

FISHMAN, R. A.: Factors influencing the exchange of sodium between plasma and cerebrospinal fluid. J. clin. Invest. 38, 1698—1708 (1959).

FISHMAN, R. A.: J. RANSOHOFF and E. F. OSSERMAN: Factors influencing the concentration gradient of protein in cerebrospinal fluid. J. clin. Invest. **37**, 1419—1428 (1958).

FORD, D. H.: The relationship of the blood-brain barrier of the rat to ^{131}I-labelled triiodothyronine. J. nerv. ment. Dis. **129**, 530—541 (1959).

FREMONT-SMITH, F., and H. S. FORBES: Intraocular and intracranial pressure. Arch. Neurol. Psychiat. (Chic.) **18**, 550—564 (1927).

FRICK, E. u. L. SCHEID-SEYDEL: Untersuchungen mit J[131]-markiertem Albumin über Austauschvorgänge zwischen Plasma und Liquor cerebrospinalis. Klin. Wschr. **36**, 66—69 (1958a).

— — Untersuchungen mit J[131]-markiertem γ-Globulin zur Frage der Abstammung der Liquoreiweißkörper. Klin. Wschr. **36**, 857—863 (1958b).

— — Untersuchungen mit J[131]-markiertem β-Globulin zur Frage der Abstammung der Liquoreiweißkörper. Klin. Wschr. **38**, 1240—1243 (1960).

GAITONDE, M. K., and D. RICHTER: The metabolic activity of the proteins of the brain. Proc. roy. Soc. B **145**, 83—99 (1956).

GEIGER, A.: Correlation of brain metabolism and function by the use of a brain perfusion method *in situ*. Physiol. Rev. **38**, 1—20 (1958).

— J. MAGNES, R. M. TAYLOR and M. VERALLI: Effect of blood constituents on uptake of glucose and on metabolic rate of the brain in perfusion experiments. Amer. J. Physiol. **177**, 138—149 (1954).

—, and S. YAMASAKI: Cytidine and uridine requirements of brain. Fed. Proc. **15**, 71 (1956).

GEINERT, F., u. H. MATIAR: Das elektrophoretische Proteinspektrum in verschiedener Höhe des Liquorsystems. Dtsch. Z. Nervenheilk. **179**, 111—119 (1959).

GERSCHENFELD, H. M., F. WALD, J. A. ZADUNAISKY and E. D. P. DE ROBERTIS: Function of astroglia in the water-ion metabolism of the central nervous system. An electron microscope study. Neurology (Minneap.) **9**, 412—425 (1959).

GOLDBERG, M. A., C. F. BARLOW and L. J. ROTH: The effects of carbon dioxide on the entry and accumulation of drugs in the central nervous system. J. Pharmacol. exp. Ther. **131**, 308—318 (1961).

GRAZER, F. M., and C. D. CLEMENTE: Developing blood-brain barrier to trypan blue. Proc. Soc. exp. Biol. (N.Y.) **94**, 758—760 (1957).

GREEN, J. B., and M. A. OLDEWURTEL: Cerebrospinal fluid enzyme studies. J. Neurosurg. **17**, 70—74 (1960).

GREIG, M. E., and A. J. GIBBONS: Effect of various psychomimetic drugs on rate of appearance of carbon-14 in the tissues of mice after administration of C[14] glucose. Amer. J. Physiol. **196**, 803—806 (1959).

GRÖNTOFT, O.: Intracranial haemorrhage and blood-brain barrier problems in the newborn. Acta path. microbiol. scand., Suppl. C, pp. 109 (1954).

GRUHN, I.: Untersuchungen über die Barrierenfunktion der Plazenta mit radioaktivem Phosphor unter normalen Bedingungen und bei der experimentellen Epilepsie. Zbl. ges. Gynäk. Geburtsh. **79**, 1025—1035 (1957).

HAIN, R. F., and J. NUTTER: Cerebrospinal fluid enzymes as a function of age. Arch. Neurol. (Chic.) **2**, 331—337 (1960).

HARREVELD, A. VAN: Asphyxial changes in the cerebellar cortex. J. cell. comp. Physiol. **57**, 101—110 (1961).

— N. K. HOOPER and J. T. CUSICK: Brain electrolytes and cortical impedance. Amer. J. Physiol. **201**, 139—143 (1961).

—, and S. OCHS: Cerebral impedance changes after circulatory arrest. Amer. J. Physiol. **187**, 180—192 (1956).

—, and J. P. SCHADE: On the distribution and movements of water and electrolytes in the cerebral cortex. In: Structure and function of the cerebral cortex. Proc. 2nd Internat. Meeting Neurobiologists, Amsterdam 1959, pp. 239—256. Amsterdam: Elsevier 1960.

HARRIS, W. H., and J. A. BEAUCHEMIN: Cerebrospinal fluid calcium, magnesium, and their ratio in psychoses of organic and functional origin. Yale J. Biol. Med. **29**, 117—124 (1956).

—, and E. H. SONNENBLICK: A study of calcium and magnesium in the cerebrospinal fluid. Yale J. Biol. Med. **27**, 297—303 (1955).

HERLIN, L.: The existence of a barrier between the cerebrospinal fluid and the boundary of the brain. In: The cerebrospinal fluid. Ciba Symp., pp. 209—229. London: Churchill 1958.

HILL, N. C., N. P. GOLDSTEIN, B. F. McKENZIE, W. F. McGUCKIN and H. J. SVIEN: Cerebrospinal fluid proteins, glycoproteins and lipoproteins in obstructive lesions of the central nervous system. Brain **82**, 581—593 (1959).

— B. F. McKENZIE, W. F. McGUCKIN, N. P. GOLDSTEIN and H. J. SVIEN: Proteins, glycoproteins and lipoproteins in the serum and cerebrospinal fluid of healthy subjects. Proc. Mayo Clin. **33**, 686—698 (1958).

HIMWICH, W. A., J. C. PETERSEN and M. L. ALLEN: Hematoencephalic exchange as a function of age. Neurology (Minneap.) **7**, 705—710 (1957).

HOGBEN, C. A. M., P. WISTRAND and T. H. MAREN: Role of active transport in formation of dogfish cerebrospinal fluid. Amer. J. Physiol. **199**, 124—126 (1960).

HORSTMANN, E., u. H. MEVES: Die Feinstruktur des molekularen Rindengraues und ihre physiologische Bedeutung. Z. Zellforsch. **49**, 569—604 (1959).

HUNTER, G., and H. V. SMITH: Calcium and magnesium in human cerebrospinal fluid. Nature (Lond.) **186**, 161—162 (1960).

— — and L. M. TAYLOR: On the bromide test of permeability of the barrier between blood and cerebrospinal fluid — an assessment. Biochem. J. **56**, 588—597 (1954).

ISHIBASHI, T.: Studies on the dynamics of the cerebrospinal fluid using radioactive isotopes. II. The circulation of the cerebrospinal fluid. Tohoku J. exp. Med. **70**, 59—66 (1959).

ISHII, S., R. HAYNER, W. A. KELLY and J. P. EVANS: Studies of cerebral swelling. II. J. Neurosurg. **16**, 152—166 (1959).

JAVID, M., and P. SETTLAGE: Effect of urea on cerebrospinal fluid pressure in human subjects. J. Amer. med. Ass. **160**, 943—949 (1956).

KARCHER, D., A. LOWENTHAL et M. VAN SANDE: Determinations de la tenure du liquide céphalo-rachidien en Ca, en K et en Na. Rev. belge Path. **26**, 49—60 (1957).

— M. VAN SANDE and A. LOWENTHAL: Micro-electrophoresis in agar gel of proteins of the cerebrospinal fluid and central nervous system. J. Neurochem. **4**, 135—140 (1959).

KISSEN, A. T., and J. M. B. BLOODWORTH: Ultrastructure of retinal capillaries of the rat. Exp. Eye Res. **1**, 1—4 (1961).

KISTER, S. J.: Carbonic anhydrase inhibition. VI. The effect of acetazolamide on cerebrospinal fluid flow. J. Pharmacol. exp. Ther. **117**, 402—405 (1956).

KIYOTA, K.: Electrophoretic protein fractions and the hydrophilic property of brain tissue. I. The proteins of normal brain. J. Neurochem. **4**, 202—208 (1959a).

— Electrophoretic protein fractions and the hydrophilic property of brain tissue. II. The proteins of brains with oedema. J. Neurochem. **4**, 209—216 (1959b).

KLATZO, I., and J. MIQUEL: Observations on pinocytosis in nervous tissue. J. Neuropath. exp. Neurol. **19**, 475—487 (1960).

— — C. TOBIAS and W. HAYMAKER: Effects of alpha particle radiation on the rat brain, including vascular permeability and glycogen studies. J. Neuropath. exp. Neurol. **20**, 459—483 (1961).

— A. PIRAUX and E. J. LASKOWSKI: The relationship between oedema, blood-brain barrier and tissue elements in a local brain injury. J. Neuropath. exp. Neurol. **17**, 548—564 (1958).

KLEEMAN, C. R., H. DAVSON and E. LEVIN: Urea transport in the central nervous system. Amer. J. Physiol. **203**, 739—747 (1962).

KNAUFF, H. G., W. MIATKOWSKY u. H. ZICKGRAF: Über die freien Aminosäuren des Liquor cerebrospinalis und ihren Nachweis mit kombinierten papierchromato- graphischen und elektropherographischen Methoden. Z. klin. Med. **155**, 483—505 (1959).

KNOPP, L. M., J. R. ATKINSON and A. A. WARD: Effect of diamox on cerebrospinal fluid pressure of cat and monkey. Neurology (Minneap.) **7**, 119—123 (1957).

KOBAYASHI, O., and B. KODAMA: Studies on electrolytes in the cerebrospinal fluid in children. Mie med. J. **5**, 97—107 (1955).

KOENIG, H.: An autoradiographic study of nucleic acid and protein turnover in the mammalian neuraxis. J. biophys. biochem. Cytol. **4**, 785—792 (1958).

KULENKAMPFF, H.: Das Verhalten der Neuroglia in den Vorderhörnern des Rückenmarks der weißen Maus unter dem Reiz physiologischer Tätigkeit. Z. Anat. Entwickl.-Gesch. **116**, 304—312 (1952).

LAJTHA, A.: The development of the blood-brain barrier. J. Neurochem. **1**, 216—227 (1957).

— Amino acid and protein metabolism of brain. II. The uptake of L-lysine by brain and other organs of the mouse at different ages. J. Neurochem. **2**, 209—215 (1958).

— Turnover of leucine in mouse tissues. J. Neurochem. **3**, 358—365 (1959).

— S. BERL and H. WAELSCH: Amino acid and protein metabolism of the brain. IV. The metabolism of glutamic acid. J. Neurochem. **3**, 322—332 (1959).

— S. FURST, A. GERSTEIN and H. WAELSCH: Amino acid and protein metabolism of the brain. I. Turnover of free and protein bound lysine in brain and other organs. J. Neurochem. **1**, 289—300 (1957).

—, and P. MELA: The brain barrier system. I. The exchange of free amino acids between plasma and brain. J. Neurochem. **7**, 210—217 (1961).

LANZA, A., e G. LOMBARDO: Sulle frazioni proteiche liquorale del bambino normale. Boll. Soc. ital. Biol. sper. **35**, 1561—1563 (1959).

LAURENCE, K. M.: Some applications of the urinary phenol-sulphonphthalein excretion test in hydrocephalus and related conditions. Brain **82**, 551—565 (1959).

LEE, J. C.: Effect of air embolism on permeability of cerebral blood vessels. Neurology (Minneap.) **9**, 619—625 (1959).

—, and J. OLSZEWSKI: Permeability of cerebral blood vessels in healing of brain wounds. Neurology (Minneap.) **9**, 7—14 (1959).

— — Penetration of radioactive bovine albumin from cerebrospinal fluid into brain tissue. Neurology (Minneap.) **10**, 814—822 (1960).

— — Increased cerebrovascular permeability after repeated electroshocks. Neurology (Minneap.) **11**, 515—519 (1961).

LENDING, M., L. B. SLOBODY and J. MESTERN: Effect of hyperoxia, and hypoxia on blood-cerebrospinal fluid barrier. Amer. J. Physiol. (Lond.) **200**, 959—962 (1961).

LEUSEN, I.: Influence de la teneur en calcium et en potassium du liquide céphalo-rachidien sur le système vasomoteur. Arch. int. Pharmacodyn. **75**, 422—424 (1948).

— Au sujet de la spécificité de l'action du CO_2 sur le centre respiratoire. Arch. int. Physiol. **57**, 456—458 (1950).

LLOYD, B. B., and K. B. TAYLOR: Genesis of human cerebrospinal fluid. J. appl. Physiol. **14**, 401—404 (1959).

LOCKSLEY, H. B.: Circulatory and cerebrospinal fluid adjustments to experimentally regulated intracranial pressures. Fed. Proc. **17**, 390 (1958).

LOESCHCKE, H. H., u. H. P. KOEPCHEN: Über das Verhalten der Atmung und des arteriellen Drucks bei Einbringen von Veratridin, Lobelin und Cyanid in den Liquor cerebrospinalis. Pflügers Arch. ges. Physiol. **266**, 586—610 (1958a).

— — Beeinflussung von Atmung und Vasomotorik durch Einbringen von Novocain in die Liquorräume. Pflügers Arch. ges. Physiol. **266**, 611—627 (1958b).

— — u. K. H. GERTZ: Über den Einfluß von Wasserstoffionenkonzentration und CO_2-Druck im Liquor cerebrospinalis auf die Atmung. Pflügers Arch. ges. Physiol. **266**, 569—585 (1958).

LOGOTHETIS, J.: Free amino acid content of cerebrospinal fluid in humans and dogs. Neurology (Minneap.) **8**, 299—302 (1958).

LOURIE, H., W. J. WEINSTEIN and J. L. O'LEARY: Effect of hypothermia upon vital staining of the brain. J. nerv. ment. Dis. **130**, 1—5 (1960).

LUMSDEN, C. E.: Observations on the choroid plexus maintained as an organ in tissue culture. In: The cerebrospinal fluid. Ciba Symp., pp. 97—123. London: Churchill 1958.

—, and C. M. POMERAT: Normal oligodendrocytes in tissue culture. Exp. Cell Res. **2**, 103—114 (1951).

LUSE, S.: Electron microscopic observations of the nervous system. J. biophys. biochem. Cytol. **2**, 531—542 (1956).

—, and B. HARRIS: Electron microscopy of the brain in experimental oedema. J. Neurosurg. **17**, 439—446 (1960).

— — Brain ultrastructure in hydration and dehydration. Arch. Neurol. (Chic.) **4**, 139—152 (1961).

MAGEE, P. N., H. B. STONER and J. M. BARNES: The experimental production of oedema in the central nervous system of the rat by triethyltin compounds. J. Path. Bact. **73**, 107—124 (1957).

MAGNES, J., and S. HESTRIN-LERNER: Effect of 5-hydroxytryptophan and serotonin on glucose content and functional activity of the perfused cat brain. J. Neurochem. **5**, 128—134 (1959).

MALAMANI, V., e F. CAMPAGNARI: Osservazioni sulla pressione intracerebrale. Boll. Soc. ital. Biol. sper. **32**, 1041—1043 (1956).

MANERY, J. F., and A. B. HASTINGS: The distribution of electrolytes in mammalian tissues. J. biol. Chem. **127**, 657—676 (1939).

MAREN, T. H., and F. J. FISCHER: Abolition of the plasma-cerebrospinal fluid chloride gradient by acetazolamide in c.s.f. secretion. Fed. Proc. **18**, 1654 (1959).

—, and B. ROBINSON: The pharmacology of acetazolamide as related to cerebrospinal fluid and the treatment of hydrocephalus. Johns Hopk. Hosp. Bull. **106**, 1—24 (1960).

MARKS, V.: True glucose content of lumbar and ventricular cerebrospinal fluid. J. clin. Path. **13**, 82—84 (1960).

MAXWELL, D. S., and D. C. PEASE: The electron microscopy of the choroid plexus. J. biophys. biochem. Cytol. **2**, 467—474 (1956).

MAYER, S., R. P. MAICKEL and B. B. BRODIE: Kinetics of penetration of drugs and other foreign compounds into cerebrospinal fluid and brain. J. Pharmacol. **127**, 205—211 (1959).

MAYNARD, E. A., R. L. SCHULTZ and D. C. PEASE: Electron microscopy of the vascular bed of the rat cerebral cortex. Amer. J. Anat. **100**, 409—433 (1957).

MILLEN, J. W., and A. HESS: The blood-brain barrier: an experimental study with vital dyes. Brain **81**, 248—257 (1958).

—, and D. H. M. WOOLLAM: Vitamins and the cerebrospinal fluid. In: The cerebrospinal fluid. Ciba Symp., pp. 168—188. London: Churchill 1958.

MORRISON, A. B.: The distribution of intravenously injected inulin in the fluids of the nervous system of the dog and rat. J. clin. Invest. **38**, 1769—1777 (1959).

MÜTING, D., u. K. N. SHIVARAM: Quantitative papierchromatographische Bestimmung der freien Aminosäuren im Liquor cerebrospinalis gesunder Menschen. Hoppe-Seylers Z. physiol. Chem. **317**, 34—38 (1960).

NIESSING, K.: Zellreaktionen der Makroglia bei Narkose. Z. mikr.-anat. Forsch. **56**, 173—189 (1950).

NIXON, D. A.: Inositol concentration in the cerebrospinal and ocular fluids and tissues of the foetal and adult sheep. Nature (Lond.) **184**, 906—907 (1959).

OKSCHE, A.: Histologische Untersuchungen über die Bedeutung des Ependyms, der Glia und der Plexus chorioidei für den Kohlenhydratstoffwechsel. Z. Zellforsch. **48**, 76—129 (1958).

Olsen, N. S., and G. G. Rudolph: Transfer of sodium and bromide ions between blood, cerebrospinal fluid and brain tissue. Amer. J. Physiol. **183**, 427—432 (1955).

Padmavati, S., Y. K. Bakshi, G. V. A. Pantulu and B. Singh: Separation and identification of sugars in cerebrospinal fluid by paper chromatography. Neurology (Minneap.) **6**, 20—24 (1958).

Palay, S. L.: An electron microscopical study of neuroglia. In: Biology of neuroglia, edit. Windle, pp. 23—38. Springfield: Thomas 1958.

Pappenheimer, J. R.: Passage of molecules through capillary walls. Physiol. Rev. **33**, 387—423 (1953).

— S. R. Heisey and E. F. Jordan: Active transport of Diodrast and phenolsulfonphthalein from cerebrospinal fluid to blood. Amer. J. Physiol. **200**, 1—10 (1961).

Pappius, H. M., and K. A. C. Elliott: Water distribution in incubated slices of brain and other tissues. Canad. J. Biochem. **34**, 1007—1022 (1956).

Paupe, J.: Comparaison entre fractions calciques de liquides céphalo-rachidiens et de sérums normaux chex l'homme. C. R. Soc. Biol. (Paris) **151**, 318—320 (1957).

Pease, D. C.: Infolded basal plasma membranes found in epithelia noted for their water transport. J. biophys. biochem. Cytol. **2**, Suppl. 203—208 (1956).

Petersdorf, R. G., D. R. Swamer and M. Garcia: Synergistic action of pneumococci and leukocytes in lowering cerebrospinal fluid glucose. Proc. Soc. exp. Biol. (N.Y.) **103**, 380—382 (1960a).

— — — Further observations on mechanism of hypoglycorrhachia in experimental meningitis. Proc. Soc. exp. Biol. (N.Y.) **104**, 65—68 (1960b).

Pilliero, S. J., and M. Lending: Protein studies in normal newborn infants. J. Dis. Child. **97**, 785—789 (1959).

Prockop, L. D., L. S. Schanker and B. B. Brodie: Passage of saccharides from cerebrospinal fluid to blood. Science **134**, 1424 (1961).

Rall, D. P., E. Moore, N. Taylor and C. G. Zubrod: The blood-cerebrospinal fluid barrier in man. Arch. Neurol. (Chic.) **4**, 318—322 (1961).

— J. R. Stabenau and C. G. Zubrod: Distribution of drugs between blood and cerebrospinal fluid; general methodology and effect of p_H gradients. J. Pharmacol. exp. Ther. **125**, 185—193 (1959).

—, and C. G. Zubrod: Permeability of the blood-cerebrospinal fluid barrier in the dogfish to sulfanilic acid and p-aminohippurate. Fed. Proc. **19**, 80 (1960).

Rao, B. S. S. R., and M. V. Govindaswamy: Studies on the cerebrospinal fluid. I. Amino acid composition. J. All-India Inst. ment. Hlth **1**, 47—59 (1958).

Reed, D. J., and D. M. Woodbury: Kinetics of C^{14}-sucrose distribution in cerebral cortex, cerebrospinal fluid, and plasma of rats. Fed. Proc. **19**, 80 (1960).

Ringerty, N. R.: On the sulphate metabolism of the mouse brain. Exp. Cell Res. **10**, 230—233 (1956).

Robertis, E. de, H. M. Gerschenfeld and F. Wald: Cellular mechanism of myelination in the central nervous system. J. biophys. biochem. Cytol. **4**, 651—658 (1958).

Roboz, E., R. R. Apostol, A. M. Luft and W. C. Hess: Determination of the carbohydrates in the glycoproteins and mucoproteins of cerebrospinal fluid. J. Lab. clin. Med. **49**, 796—801 (1957).

— J. B. Murphy, W. C. Hess and F. M. Foster: Isolation and determination of the glycoproteins in cerebrospinal fluid. Proc. Soc. exp. Biol. (N.Y.) **89**, 691—694 (1955).

Rodriguez, L. O.: Experiments on the histologic locus of the hemato-encephalic barrier. J. comp. Neurol. **102**, 27—39 (1955).

Roth, L. J., J. C. Schoolar and C. F. Barlow: Sulphur-35 labelled acetazolamide in cat brain. J. Pharmacol. exp. Ther. **125**, 128—136 (1959).

Rougemont, J. de, A. Ames, F. B. Nesbitt and H. F. Hofmann: Fluid formed by choroid plexus. J. Neurophysiol. **23**, 485—495 (1960).

Royer, P.: Perfusion endocavitaire cérébrale. Biol. méd. (Paris) **39**, 237—269 (1950).

Samachson, J., B. Kabakow, H. Spencer and D. Laszlo: Comparative passage of calcium and strontium from plasma into cerebrospinal fluid in man. Proc. Soc. exp. Biol. (N.Y.) **102**, 287—290 (1959).

Samorajski, T., and R. A. Moody: Changes in the blood-brain barrier after exposure of the brain. Arch. Neurol. Psychiat. (Chic.) **78**, 369—376 (1957).

Scheibel, M. E., and A. B. Scheibel: Neurons and neuroglia cells as seen with the light microscope. In: Biology of neuroglia, edit. Windle, pp. 5—23. Springfield: Thomas 1958.

Scheiffarth, F., H. Götz, G. Berg u. H. Hopfensperger: Immuno-elektrophoretische Studien am normalen und pathologischen Liquor. Klin. Wschr. **36**, 678—681 (1958).

Schoolar, J. C., C. F. Barlow and L. J. Roth: The penetration of carbon-14 urea into cerebrospinal fluid and various areas of the cat brain. J. Neuropath. exp. Neurol. **19**, 216—227 (1960).

Schultz, R. L., E. A. Maynard and D. C. Pease: Electron microscopy of neurons and neuroglia of cerebral cortex and corpus callosum. Amer. J. Anat. **100**, 369—407 (1957).

Schultze, H. E., M. Schönenberger u. G. Schwick: Über ein Präalbumin des menschlichen Serums. Biochem. Z. **328**, 267—284 (1956).

Sibrik, I. de, and D. S. O'Doherty: Phosphatases and phospholipases in the central nervous system. Arch. Neurol. (Chic.) **2**, 537—546 (1960).

Silverstein, A., and E. M. Greenspan: Mucoprotein content of cerebrospinal fluid. Confin. neurol. (Basel) **19**, 306—318 (1959).

Sjöstrand, F. A., and J. Rhodin: The ultrastructure of the proximal convoluted tubules of the mouse kidney as revealed by high resolution electron microscopy. Exp. Cell Res. **4**, 426—456 (1953).

Slobody, L. B., D. C. Yang, M. Lending, F. J. Borrelli and M. Tyree: Effect of severe hypoxia on blood-cerebrospinal fluid barrier. Amer. J. Physiol. **190**, 365—370 (1957).

Small, H. S., S. W. Weitzner and G. G. Nahas: Spinal fluid pressure changes during apneic oxygenation and carbon dioxide administration. Fed. Proc. **17**, 593 (1958).

Smythe, L., G. Smythe and P. Settlage: The effect of intravenous urea on cerebrospinal fluid pressure in monkeys. J. Neuropath. exp. Neurol. **9**, 438—442 (1950).

Spina-França, A.: Electroforese em papel das proteinas do liquido cefalorraquidiano: IV. Valores normais. Arch. Neuro-psiquiat. (S. Paulo) **18**, 19—28 (1960).

Sporn, M. B., W. Dingman, A. Defalco and R. K. Davies: The synthesis of urea in the living rat brain. J. Neurochem. **5**, 62—67 (1959).

Stabenau, J. R., K. S. Warren and D. P. Rall: The role of p_H gradient in the distribution of ammonia between blood and cerebrospinal fluid, brain and muscle. J. clin. Invest. **38**, 373—383 (1959).

Steger, J.: Elektrophoretische Untersuchungen des Liquors. Dtsch. Z. Nervenheilk. **171**, 1—19 (1953).

Stein, W. H., and S. Moore: The free amino acids of human blood plasma. J. biol. Chem. **211**, 915—926 (1954).

Stern, L., et R. Peyrot: Le fonctionnement de la barrière hémato-encéphalique aux diverses stades de développement chez diverses espèces animales. C. R. Soc. Biol. (Paris) **96**, 1124—1126 (1927).

Stern, W. E.: Studies in experimental brain swelling and brain compression. J. Neurosurg. **16**, 676—698 (1959).

Streicher, E.: The entry of $P^{32}O_4$ into rat brain. J. cell. comp. Physiol. **48**, 139—146 (1956).

— Thiocyanate space of rat brain. Amer. J. Physiol. **201**, 334—336 (1961).

— D. P. Rall and J. R. Gaskins: The distribution of thiocyanate between the cerebrospinal fluid and plasma. Fed. Proc. **20**, 392 (1961).

STUBBS, J., and J. PENNYBACKER: Reduction of intracranial pressure with hypertonic urea. Lancet 1960 I, 1094—1097.

SWEET, W. H., G. L. BROWNELL, J. A. SCHOLL, D. R. BOWSHER, P. BENDA and E. E. STICKLEY: The formation, flow and absorption of cerebrospinal fluid; newer concepts based on studies with isotopes. Ass. Res. nerv. Dis. Proc. 34, 101—159 (1954).

— B. SELVERSTONE, A. SOLOMON and L. BAKAY: Studies of formation, diffusion and absorption of constituents of cerebrospinal fluid in man. J. clin. Invest. 28, 814 (1949).

TORACK, R. M., M. BESEN and N. H. BECKER: Localization of adenosinetriphosphatase in capillaries of the brain as revealed by electron microscopy. Neurology (Minneap.) 11, 71—76 (1961).

— R. D. TERRY and H. M. ZIMMERMAN: The fine structure of cerebral fluid accumulation. I. Swelling secondary to cold injury. Amer. J. Path. 35 1135—1147 (1959).

— — — The fine structure of cerebral fluid accumulation. II. Swelling produced by triethyl tin poisoning and its comparison with that in the human brain. Amer. J. Path. 36 273—287 (1960).

TSCHIRGI, R. D.: Protein complexes and the impermeability of the blood-brain barrier to dyes. Amer. J. Physiol. 163, 756 P (1950).

— Chemical environment of the central nervous system. In: Handbook of Physiology, vol. III, pp. 1865—1890, edit. FIELD, MAGOUN and HALL. Baltimore: Amer. Physiol. Soc. 1960.

— R. W. FROST and J. L. TAYLOR: Inhibition of cerebrospinal fluid formation by a carbonic anhydrase inhibitor (Diamox). Proc. Soc. exp. Biol. (N.Y.) 87, 373—376 (1954).

—, and J. L. TAYLOR: Slowly changing bioelectric potentials associated with the blood-brain barrier. Amer. J. Physiol. 195, 7—22 (1958).

TUBIANA, M., P. BENDA and J. CONSTANS: Sodium radioactif (^{24}Na) et liquide céphalo-rachidien. Rev. neurol. 85, 17—35 (1951).

TURNER, L.: The structure and relationships of arachnoid granulations. In: Cerebrospinal fluid. Ciba Symp., pp. 32—54. London: Churchill 1958.

— The structure of arachnoid granulations with observations on their physiological and pathological significance. Ann. roy. Coll. Surg. Engl. 29, 237—264 (1961).

UZMAN, L. L., E. A. BERING and C. E. MORRIS: The neuraminic acid content of cerebro-spinal fluid as affected by neurological diseases. J. clin. Invest. 38, 1756—1768 (1959).

—, and M. K. RUMLEY: Neuraminic acid as a constituent of human cerebrospinal fluid. Proc. Soc. exp. Biol. (N.Y.) 93, 497—501 (1956).

VULPE, M., A. HAWKINS and B. ROZDILSKY: Permeability of cerebral blood vessels in experimental allergic encephalomyelitis studied by radioactive iodinated bovine albumin. Neurology (Minneap.) 10, 171—177 (1960).

WANG, J.: Penetration of radioactive sodium and chloride into cerebrospinal fluid and aqueous humour. J. gen. Physiol. 31, 259—268 (1948).

WART, C. A. VAN, J. R. DUPONT and L. KRAINTZ: Transfer of radioiodinated human serum albumin (RISA) from cerebrospinal fluid to blood plasma. Proc. Soc. exp. Biol. (N.Y.) 103, 708—710 (1960).

— — — Effect of acetazolamide on passage of protein from cerebrospinal fluid to plasma. Proc. Soc. exp. Biol. (N.Y.) 106, 113—114 (1961).

WEED, L. H.: The pathways of escape from the subarachnoid spaces with particular reference to the arachnoid villi. J. med. Res. 31, 51—91 (1914).

—, and P. S. MCKIBBEN: Pressure changes in the cerebrospinal fluid following intra-venous injections of solutions of various concentrations. Amer. J. Physiol. 48, 512—530 (1919).

WEIL-MALHERBE, H., J. AXELROD and R. TOMCHICK: Blood-brain barrier for adrenaline. Science 129, 1226—1227 (1959).

WEISE, H.: Untersuchungen über den Kationengehalt des Liquor cerebrospinalis. Z. ges. exp. Med. **131**, 353—358 (1959).

WELCH, K., and V. FRIEDMAN: The cerebrospinal fluid valves. Brain **83**, 454—469 (1960).

WHITE, J. C., J. R. BROOKS, J. C. GOLDTHWAIT and R. D. ADAMS: Changes in brain volume and blood content after experimental concussion. Ann. Surg. **118**, 619—638 (1943).

WINDLE, W. F.: The biology of neuroglia. Springfield: Thomas 1958.

WINZLER, R. J.: Determination of serum glycoproteins. In: Methods of biochemical analysis, vol. 2, edit. GLICK. New York: Interscience 1955.

WISLOCKI, G. B., and A. J. LADMAN: The fine structure of the mammalian choroid plexus. In: The cerebrospinal fluid, Ciba Symp., pp. 55—79. London: Churchill 1958.

WOLFF, P. H., and R. D. TSCHIRGI: Inability of cerebrospinal fluid to nourish the spinal cord. Amer. J. Physiol. **184**, 220—222 (1956).

WOODBURY, D. M., P. S. TIMIRAS, A. KOCH and A. BALLARD: Distribution of radio-chloride, radiosulphate and inulin in brain of rats. Fed. Proc. **15**, 501—502 (1956).

WOOLLAM, D. H. M., and J. W. MILLEN: The relationship between hypovitaminosis A and the cerebrospinal-fluid pressure in the chick: an experimental study. Brit. J. Nutr. **10**, 355—363 (1956).

WOOLLEY, D. W., and E. N. SHAW: Evidence for the participation of serotonin in mental processes. Ann. N.Y. Acad. Sci. **66**, 649—667 (1957).

WORATZ, G., u. W. ROTZSCH: Die statistische Verteilung von Natrium, Kalium und Calcium im Liquor. Dtsch. Z. Nervenheilk. **181**, 252—260 (1960).

WYCKOFF, R. W. G., and J. Z. YOUNG: The motorneuron surface. Proc. roy. Soc. B **144**, 440—450 (1956).

ZUBROD, C. G., and D. P. RALL: Distribution of drugs between blood and cerebrospinal fluid in various vertebrate classes. J. Pharmacol. exp. Ther. **125**, 185—193 (1959).

Vagal Afferent Fibres

By

A. S. PAINTAL[*]

With 36 Figures

Contents

---

[*] From the Department of Physiology, All-India Institute of Medical Sciences, Ansari Nagar, New Delhi-16, India.

I. Introduction

The cervical vagus consists of about 30 thousand fibres of which about 24000 are sensory in function (AGOSTONI, CHINNOCK, DALY and MURRAY 1957). Of the latter about 3 thousand are myelinated and have been the centre of attraction in electrophysiological studies chiefly owing to the relative ease with which impulses can be recorded in them. In the 19th International Physiological Congress, WHITTERIDGE reviewed briefly all that was known about these myelinated afferent fibres which arose from endings of various kinds in the heart and lungs (WHITTERIDGE 1953) and although more detailed information about certain myelinated sensory endings has been gained since that time, the main advance has been in the direction of recording impulse in non-medullated fibres from endings in the gastro-intestinal tract [PAINTAL 1953 (d), 1954 (a), (b); IGGO 1955, 1957 (b)]. Till IGGO's clear demonstration (IGGO 1958) that the fibres concerned are mostly non-medullated, it was generally imagined that unitary discharges in non-medullated fibres could be recorded only with great difficulty owing to their small size. It was felt that if recording impulses in single fibres of small myelinated fibres presented certain difficulties, it would be still more difficult to record impulses in non-medullated fibres. However, as pointed out by IGGO (1958), the situation is probably not comparable because non-medullated fibres occur in small bundles which can probably be dissected as such. Since there are many more non-medullated afferent fibres (21 000) than medullated ones (3 000) (Fig. 1), this therefore opens up a practically unexplored field for future work.

The responses of some of the sensory endings connected to vagal afferent fibres were reviewed twelve years ago in this Journal by SCHAEFER (1950). Thereafter other reviews have appeared (DAWES 1953, 1954; DAWES and COMROE 1954; AVIADO and SCHMIDT 1955; HEYMANS and NEIL 1958; CHERNIGOVSKY 1960), but since in most of these reviews the emphasis has been on reflex effects, they contain only brief descriptions of the behaviour of the sensory endings or their fibres. In the present review attention will be paid only to the responses of the endings and the nature of their fibres and the reflex effects produced by them will be ignored. The responses of the endings to chemical substances will not be discussed either. These have been briefly reviewed elsewhere [PAINTAL 1956 (b); WHITTERIDGE 1958].

II. Afferent fibre composition of the vagus

The afferent and efferent fibre composition of the vagus and its branches has been worked out using the conventional degeneration techniques [JONES 1932, 1937; HEINBECKER and O'LEARY 1933; FOLEY and DuBois 1934, 1937; DuBois and FOLEY 1936; DALY and EVANS 1953; EVANS and MURRAY 1954 (a), (b); AGOSTONI et al. 1957; HOFFMAN and KUNTZ 1957; MURRAY 1957] and it has been possible to determine the number of afferent fibres of different diameters that are present in the main vagal trunk or its branches after cutting the vagus above the nodose ganglion and allowing the efferent fibres to degenerate [DALY and EVANS 1953; EVANS and MURRAY 1954 (a); AGOSTONI et al. 1957]. Although there appeared to be some conflicting results some years ago, it is now clear from the recent papers by DALY and EVANS (1953) and AGOSTONI et al. (1957) that in the cat, as also in the rabbit [EVANS and MURRAY 1954 (a), (b)] the fibres that survive after supranodose vagotomy are almost entirely sensory with their cell stations in the nodose ganglion. This confirms the main conclusions of earlier workers quoted above except HEINBECKER and O'LEARY (1933) who believed that certain motor fibres originated in the nodose ganglion.

In the case of the vagus, unlike somatic nerves, it is not possible to study efferent fibres after allowing the sensory ones to degenerate because owing to the arrangement of the vagal rootlets (Fig. 35) it is not possible to section sensory rootlets exclusively. There are therefore no studies of deafferented vagus nerves.

A. Composition of vagal branches

The number of medullated and non-medullated afferent fibres in the various branches of the vagus nerve are shown roughly in Fig. 1.

The vagus gives off the auricular branch immediately after the rootlets unite to form the main vagal trunk. This branch which appears to be a typical cutaneous nerve and almost exclusively sensory consists of about 6 to 8 thousand fibres of which about a quarter are myelinated. All these fibres have their cell stations in the jugular ganglion (DU BOIS and FOLEY 1936). In contrast the pharyngeal branch which is given off next is composed mostly of efferent fibres because as shown by RANSON, FOLEY and ALPERT (1933), this branch undergoes almost complete degeneration after section of the vagal rootlets. It apparently has a few small myelinated sensory fibres which arise from cells in the jugular ganglion.

The superior laryngeal nerve consists mostly of sensory fibres because about 2200 fibres out of about 2800 medullated fibres remain after intra-cranial vagotomy (DU BOIS and FOLEY 1936). In the rat there are only about 700 medullated fibres [ANDREW 1956 (a)] and in man about 15 000 (OGURA and LAM 1953). The medullated fibres range from 1.5—8 μ and of these, the 1.5—5 μ ones are more numerous (DU BOIS and FOLEY 1936). The recurrent laryngeal nerve is composed of large and small medullated fibres and a few non-medullated fibres (DU BOIS and FOLEY 1936; MURRAY 1957). The small medullated fibres ranging from 1—8 μ (MURRAY 1957) are sensory and supply the oesophagus and the trachea and some of them arise from tracheo-bronchial endings [WIDDICOMBE 1954 (b)]. These fibres number about 5 to 6 hundred and they do not degenerate after supranodose vagotomy (DU BOIS and FOLEY 1936; MURRAY 1957). The larger medullated fibres of the recurrent ranging in diameter from 4—18 μ number about 450. They have a unimodal distribution with a peak at about 10—12 μ. Nearly all these fibres which supply the internal muscles of the larynx degenerate after supranodose vagotomy (DuBOIS and FOLEY 1936; MURRAY 1957). Similar observations have been made in the rabbit [EVANS and MURRAY 1954 (a)].

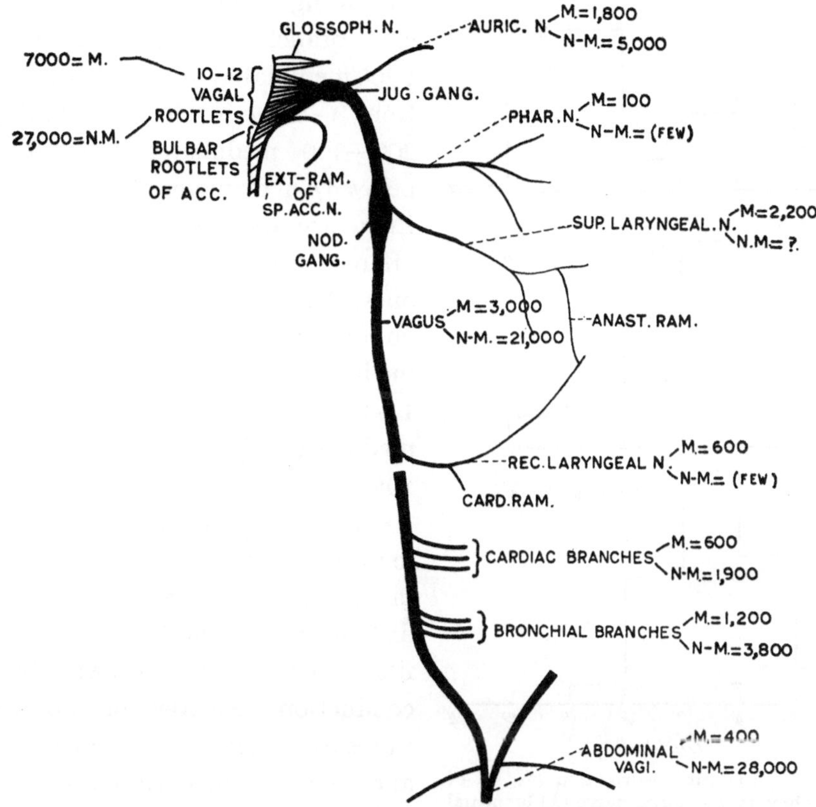

Fig. 1. Schematic diagram to show the number of medullated (M) and non-medullated (N-M) sensory fibres present in the vagus nerve and its branches. The data was obtained from DuBOIS and FOLEY (1936), DALY and EVANS (1953) and AGOSTONI et al. (1957)

It is important to keep these observations on the recurrent laryngeal nerve in mind because in the neck where impulses in vagal afferent fibres are commonly recorded, the fibres of the recurrent laryngeal form a part of the vagus and although the fibres are usually confined to a section of the cervical vagus (see Fig. 1 of plate 1 in AGOSTONI et al. 1957), it is not possible to separate out this section in electrophysiological studies. This therefore may not uncommonly result in vagal filaments being isolated, which though they contain fast conducting fibres are nevertheless totally silent. Since the maximum conduction velocity of vagal afferent fibre is 60 m/sec [PAINTAL 1953 (c)], it is justifiable to assume that any fibre with a conduction velocity greater than 60 m/sec which does not show any normal discharge and in which a discharge cannot be aroused by rapid inflation of the lungs, is an efferent fibre destined for the internal laryngeal muscles. This can be proved by stimulating the recurrent laryngeal nerve itself and looking for the appropriate evoked impulses in the filament concerned.

The composition of the bronchial and cardiac branches of the vagus has been studied chiefly by Daly and Evans (1953) and Agostoni et al. (1957).

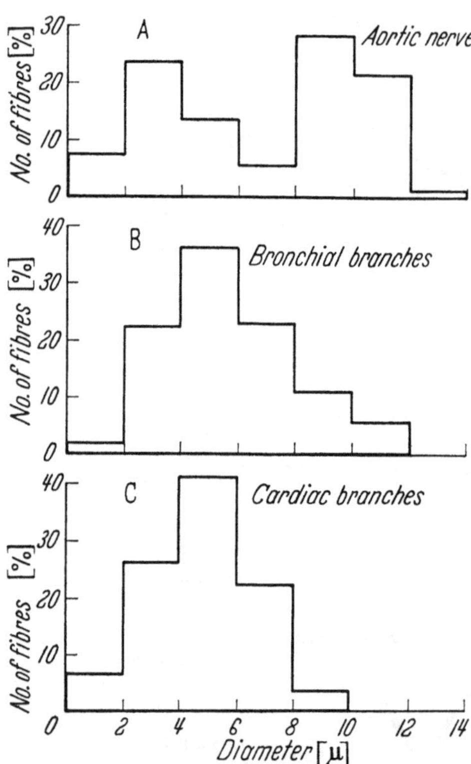

Fig. 2A—C. Diameter distribution (in %) of afferent fibres in the aortic nerve (A) bronchial (B) and cardiac (C) branches of the vagus nerve. For the aortic nerve the data was obtained from Table 3 and for the bronchial and cardiac branches from Table 6 of Agostoni et al. (1957). In A, B and C 100 % is respectively 343, 1,259 and 629 fibres

According to these studies all the bronchial branches together contain about 1 200 medullated afferent fibres ranging from 1—14 μ in diameter and about 200—300 medullated efferent fibres below 4 μ in diameter. In these branches there are about 3 800 non-medullated afferent fibres and about 800 efferent ones. All the cardiac branches together on one side contain about 500—700 medullated afferent fibres ranging from 1—12 μ (Fig. 2) and about 1 900 non-medullated fibres (Agostoni et al. 1957).

As shown in Fig. 2B and C the majority of the medullated fibres fall in the 2—8 μ diameter range. This distribution in the bronchial branches fits in with what is known about the conduction velocities of the afferent fibres from pulmonary stretch receptors and tracheobronchial receptors (Fig. 8B; Table 1). On the other hand it is difficult to reconcile the distribution of the fibres in the cardiac branches with what is known about the conduction velocities of fibres from atrial receptors and ventricular receptors because the conduction velocities of the fibres from these endings do not exceed 30 m/sec (Table 1; Fig. 6). If one assumes a conversion factor of 6 (Hursh 1939) Fig. 2C suggests that about half the afferent fibres from cardiac endings will have a conduction velocity ranging from 30—60 m/sec. It is therefore possible that most of these larger fibres arise from pulmonary endings because records of impulse activity from these cardiac branches show significant activity from pulmonary stretch receptors (Amann and Schaefer 1943; Jarisch and Zotterman 1948; Dickinson 1950).

The abdominal branches of the vagus are characterised by a paucity of medullated afferent fibres in contrast to the large number of non-medullated afferent fibres (Fig. 1). Apparently out of 31000 fibres in the anterior and posterior vagus nerves at the level of the diaphragm there are only about 400 medullated afferent fibres all of which are below $6\,\mu$ in diameter and almost entirely afferent. All the other fibres are non-medullated and 90% of them (28000) are apparently sensory because they survive after supranodose vagotomy. There are clearly adventitial fibres in this group because AGOSTONI et al. (1957) found that after chronic cervical vagotomy, about 10% of the normal number of medullated and non-medullated fibres remained. The source of these fibres remains obscure; perhaps they are of sympathetic origin. Since the right and left vagus nerves contribute an approximately equal number of fibres to the abdominal vagus nerves, the cervical vagus must contain about 14000 non-medullated, and 200 medullated afferent fibres arising from endings in the abdominal viscera.

B. Aortic nerve

According to AGOSTONI et al. (1957) the aortic nerve in the cat is probably made up entirely of sensory fibres because there is no evidence of degeneration in this nerve after supranodose vagotomy. The nerve contains about 300 medullated and about 150 non-medullated fibres. The medullated ones range from 1—$12\,\mu$ in diameter and curiously enough are bimodally distributed with peaks at 2—$4\,\mu$ and 10—$12\,\mu$ (Fig. 2A). The authors were unable to account for this distribution because it does not exist in the carotid nerve. Another marked difference from the carotid nerve is that the fibres in the 6—$8\,\mu$ group in the carotid nerve constitute only 3.5% of the total number of medullated fibres (DE CASTRO 1951). EYZAGUIRRE and UCHIZONO (1961) have recently arrived at a similar conclusion.

Fig. 2A which is a frequency distribution of the afferent fibres in the aortic nerve has been drawn from the results of AGOSTONI et al. and represents the results of two experiments (Table 3 from AGOSTONI et al. 1957). It shows that in the aortic nerve the fibres in the 10—$12\,\mu$ range make up about 1/5th of the total number of medullated fibres. This means that one out of every 5 fibres isolated from the aortic nerve for electrophysiological studies should have a conduction velocity of 60—72 m/sec if the fibre diameter: conduction velocity ratio is assumed to be 6 as in somatic nerves (HURSH 1939). This is not what has been observed in fibres isolated at random (Fig. 3C) because the mean conduction velocity of the fibres from arterial baroreceptors is 33 m/sec and the maximum conduction velocity is 53 m/sec (Table 1) [PAINTAL 1953 (c)]. This discrepancy may be due to several reasons: 1. the factor of 6 is too high for the fibres of the aortic nerve (and also the vagus); 2. the larger diameter fibres do not originate in systemic arterial baroreceptors but in other endings; and 3. the nerves examined by Agostoni et al. were not aortic nerves but some other branch of the vagus. This is possible because occasionally certain branches join the vagus in the same way as the aortic nerve but they have no afferent fibre activity from arterial baroreceptors. On the other hand, although the aortic nerve contains predominantly fibres with a cardiac rhythm, it is usual to find fibres with a respiratory rhythm presumably arising from pulmonary stretch receptors also; such fibres could account for a few of the larger diameter fibres.

Examination of the compound action potential of the cat's aortic nerve also shows an absence of the larger diameter fibres shown in Fig. 2A. Thus in Fig. 3B the maximum conduction velocity is 47 m/sec. The compound action potential corresponds to the distribution of the conduction velocities of fibres from systemic arterial baroreceptors (Fig. 3C). Using

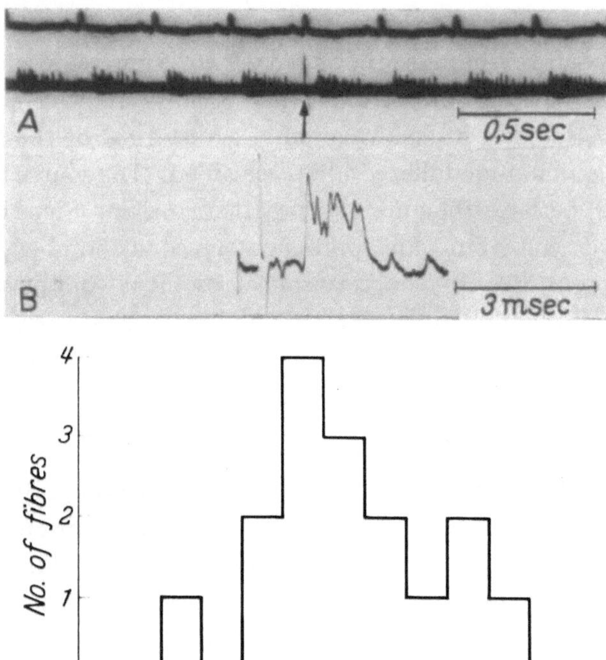

Fig. 3A—C. Conduction velocities of systemic arterial baroreceptors. A is a continuous record to show the baroreceptor discharge in the whole aortic nerve. A stimulus was applied to the vagus at arrow and the resulting compound action potential recorded in B which shows that the maximum conduction velocity of the fibres in this aortic nerve was 47 m/sec. The conduction velocities of the fibres in the main part of the potential range from 19—47 m/sec (Paintal 1962, unpublished observations). The distribution is therefore similar to that in C which shows the distribution of conduction velocities of fibres from arterial baroreceptors as determined in individual fibres [Paintal 1953 (c)]

the conversion factor of 6, Fig. 3B and C fit in with the results of De Castro (1951) and Eyzaguirre and Uchizono (1961) on the carotid nerve because 3/4 of the fibres from arterial baroreceptors have conduction velocities between 10 and 40 m/sec corresponding to fibre diameters of 2—7 μ (see also Heymans and Neil 1958). Obviously more work is needed to clear up this point along the lines of investigation elaborated by Erlanger (1928) and Erlanger and Gasser (1937) i.e. a combination of electrophysiological and histological studies followed by reconstruction of the compound action potential.

C. Cervical vagus

Of greatest interest in connection with electrophysiological studies is the composition of the cervical vagus. The composition of this part of the nerve has been studied by several investigators (see references above) but counts of fibres of different diameters have been made only by AGOSTONI et al. (1957). Of the approximately 5 000 medullated fibres in the vagus (exclusive of those in the aortic nerve) about 3 000 are afferent, and about 2 000 are efferent.

However, the great majority of the fibres are non-medullated, numbering about 25 000, and of these the large majority, about 21 000, are most probably sensory since they survive after supranodose vagotomy (AGOSTONI et al. 1957). All these afferent fibres both medullated and non-medullated have their cell bodies in the nodose ganglion and practically none in the jugular ganglion or in the few scattered ganglion cells because the number of fibres in the cervical vagus remaining after supranodose or intracranial vagotomy are essentially the same (DU BOIS and FOLEY 1936; AGOSTONI et al. 1957). All the largest medullated fibres of 12—18 μ in diameter are efferent since they degenerate after supranodose vagotomy. These fibres innervate the internal laryngeal muscles through the recurrent laryngeal nerve mentioned above (AGOSTONI et al. 1957; MURRAY 1957).

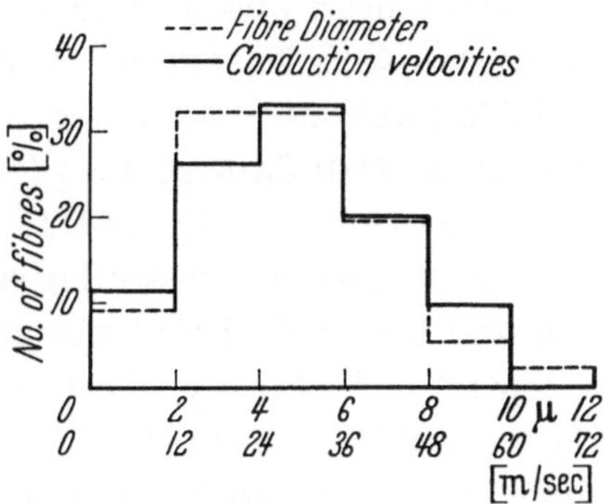

Fig. 4. Diameter distribution (in %) of afferent fibres in the cervical vagus and distribution of conduction velocities (also in %) of vagal afferent fibres. The data for diameter distribution was obtained from Table 3 of AGOSTONI et al. (1957); in this curve 100% = 2995 fibres. The data for conduction velocities were compiled from an earlier investigation [PAINTAL 1953 (c)]; 100% = 116 fibres

The great majority of medullated afferent fibres (about 87% of the total) are in the 2—8 μ range. This is shown in Fig. 4 which has been drawn from the data in Table 3 of the paper by AGOSTONI et al. (1957) and represents the average counts (expressed as a percentage of the total) from 4 vagus nerves after supranodose vagotomy. This figure excludes the few medullated fibres that are invariably left after infranodose vagotomy; these amount to an average of 75 fibres from the results of two experiments (AGOSTONI et al. 1957) and they apparently belong exclusively to the 4—6 μ group. They form about 8% of the fibres in this group. Stimulating the caudal end of the vagus in those cases where these fibres were present produced a slight rise in systemic blood pressure. This needs to be explained.

The interrupted line in Fig. 4 represents the number of fibres (in %) of each group isolated electrophysiologically and has been drawn from the data published earlier [PAINTAL 1953 (c)]. This is discussed on page 86.

III. Dissection of filaments with unitary discharges

As shown by ADRIAN and ZOTTERMAN (1926) and ADRIAN (1928) it is necessary to record sensory impulses in single active nerve fibres (unitary discharge, "single unit") in order that a proper analysis of the responses of the endings can be made. Such analysis is very difficult in preparations containing several active nerve fibres (see ADRIAN 1926; KELLER and LOESER 1926; PARTRIDGE 1933).

Various descriptions of methods for dissecting single myelinated fibres (STÄMPFLI 1952; TASAKI 1953) or filaments containing one active fibre are available [ADRIAN and BRONK 1928; MATTHEWS 1933; WHITTERIDGE 1948; DICKINSON 1950; PAINTAL 1953 (c)]. Detailed descriptions have been provided only by STÄMPFLI (1952) and TASAKI (1953) along with illustrations of each step of dissection. For dissecting single units it is only necessary to use steps C, D, E and F described by STÄMPFLI followed by section of the filament from its central end. Finally, the procedure described in his step F and G may be used profitably to subdivide the filament by gentle teasing. By this procedure it is possible to isolate many single units in one experiment and is the one employed by most investigators with slight modifications.

IV. Determination of conduction velocities

A. Methods

Several methods have been described for determining the conduction velocities of individual afferent fibres [PAINTAL 1953 (c); IGGO 1958] without the need of dissecting filaments containing only one live fibre which is an ideal to aim at. In principle the conduction velocity is determined according to conventional methods (see ERLANGER 1928) in which the time taken for the evoked impulse to travel over a fixed distance from the stimulating to the recording electrodes is determined. However, when a filament contains a few live fibres but one active fibre whose conduction velocity is to be determined, the difficulty that arises is how to establish which one of the several impulses evoked by the electrical stimulus belongs to the active fibre [PAINTAL 1953 (c)]. In this connection certain criteria to establish this point are available and are quite useful in practice (Fig. 5) [PAINTAL 1953 (c)]. It was pointed out that actual proof that a particular evoked impulse and the natural impulses arose in the same fibre could be obtained by showing that the electrically evoked impulse does not appear if it falls immediately after a preceding natural impulse owing to absolute refractoriness (Fig. 5 C). This is true if the electrical stimulus is clearly above threshold but in practice it is often not convenient to apply strong stimuli. The only really fool-proof method is to show that the electrically evoked impulse is clearly reduced in size (Fig. 5 B) or appears with increased latency (Fig. 5 A) owing to relative refractoriness if it appears within a few milliseconds after a natural impulse (Fig. 5 A and B). This is not difficult in the case of cardiovascular afferent fibres because some part of the ECG can trigger the sweep and the stimulator adjusted so that the evoked impulse appears during the relative refractory phase of natural impulse. It is also simple if the conduction time in the afferent fibre between the stimulating and recording electrodes is short because it is easy to arrange that a natural impulse triggers both the sweep and the stimulator so that the electrically evoked impulse appears during the relative refractory period of the natural impulse. With this arrangement it is convenient to have a gating circuit so that only certain natural impulses trigger the stimulator in order to avoid tetanic stimulation of the nerve when there is a high frequency discharge of impulses in the fibre. However, there are many slowly conducting fibres so that the conduction time is long and this method cannot therefore be used. In such cases one can use a suitable rate of stimulation and record superimposed sweeps till one of the evoked impulses falls in the relative

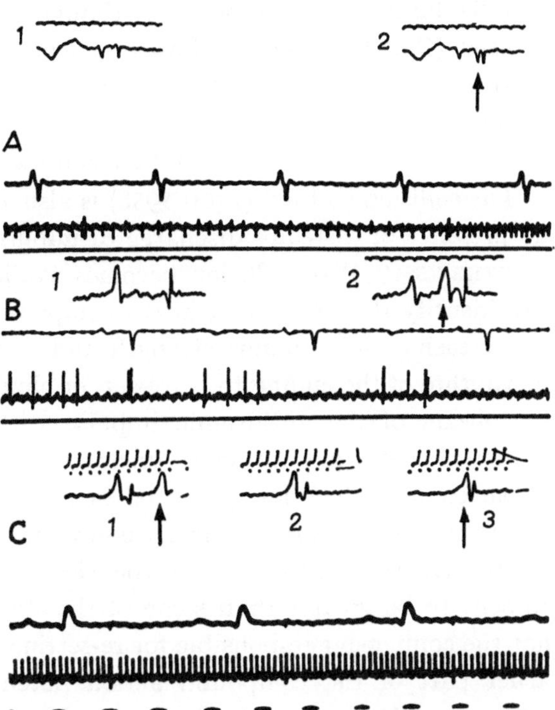

Fig. 5 A—C. The effect of setting up evoked impulses during the refractory period after impulses due to receptor activity. A and C, records from pulmonary stretch receptors, B is a record from a right atrial receptor type B. In each record the two upper traces, *to be read from right to left,* are of time in milliseconds and of action potentials for conduction time determinations. The two lower traces are of the e.c.g. and of action potentials in a single unit, and read from left to right. In C, time in $^1/_{10}$ sec. In A, sweep 2, the evoked impulse falls late in the relative refractory period due to an impulse set up by the receptor, and its conduction rate is reduced from 31—25 m/sec. In B, sweep 2, the evoked impulse is slowed and reduced in size by the impulse from the receptor. In C, sweep 3, the evoked impulse is absent in the third sweep as it has fallen in the absolute refractory period. All arrows indicate position of impulses set up by receptor activity [PAINTAL 1953 (c)]

refractory period of a randomly occurring natural impulse. Alternatively, continuous records can be taken along with sweeps as described earlier (Fig. 5).

Short of actual proof of identity of an evoked impulse with natural impulses, there are some criteria that can be used to provide strongly suggestive evidence of identity of the two. Thus one could use the shape, size and duration of the action potentials as useful criteria so that if the evoked impulse is "all or none" and is similar in size, shape and duration to the natural impulses, one could be reasonably certain that they both belonged to the same afferent fibre.

In the case of fibres in which only few impulses can be produced by natural stimuli, the method of repetitive stimulation with freely running sweeps can be used (PAINTAL 1960). However, repetitive stimulation is not desirable if atropine cannot be injected for some reason and is therefore used as a last resort.

B. Collision method

The collision method (IGGO 1958) is also very useful. This method requires the presence of a steady discharge of impulses so that evidence of re-setting (MATTHEWS 1933) or collision becomes available. The aim of this procedure is to identify the component in the compound action potential which appears on increasing the stimulus strength and which simultaneously either causes the rhythm of the ending to be re-set, or eliminates one natural impulse owing to collision of the antidromic impulse with the missing orthodromic one. However, there may be some uncertainty in identifying which component of the compound action potential is responsible for re-setting or collision, if there are several fibres or there are slow and fast components in the compound action potential particularly if the electrical threshold of some fast fibres happens to be greater than some of the slower ones. Thus it is conceivable that the component responsible for re-setting or collision may be hidden in the earlier part of the compound action potential and therefore re-setting or collision may be erroneously ascribed to the simultaneous appearance of another component in the slower part of the compound action potential. However, in spite of its limitations the method is useful in practice especially in the case of C fibres; it is not so useful in the case of fast fibres, especially when $f \ll v/2d$ where $f =$ the frequency of discharge, $v =$ conduction velocity of the fibre and $d =$ the distance between the stimulating electrodes and the ending [IGGO 1958; see also DOUGLAS and RITCHIE 1957 (a), (b)].

Finally as demonstrated by IGGO (1958) stimulating the afferent fibre somewhere close to the ending can be valuable for establishing the conduction velocity of individual afferent fibres. The procedure has been described in detail by IGGO (1958) and is particularly useful in the case of gastro-intestinal fibres. Probably it can also be used in the case of pulmonary and cardiovascular afferent fibres because it is also possible to stimulate these fibres somewhere near the ending [PAINTAL 1962]. However, it is necessary to open the chest to accomplish this. Besides, the intrathoracic conduction velocities of many fibres are lower than their respective cervical conduction velocities (p. 87).

C. Errors in measurements of conduction velocities

Owing to variable setting-up time at the stimulating electrodes (BLAIR and ERLANGER 1936), and the impulse not arising at the cathode, the use of a single conduction distance is not recommended for precise measurements of conduction velocity. Two or three conduction distances are therefore used (e.g. SANDERS and WHITTERIDGE 1946; KIRALY and KRNJEVIĆ 1959). However, in the case of the vagus, IGGO's results indicate that two conduction distances cannot be of additional value because apparently the conduction velocities in the distal parts of afferent fibres may be less than in the proximal parts (IGGO 1958). IGGO's results also suggest (see Table 1 of his paper) that if an arbitrary allowance of 0.1 msec for shock-response delay is made, this

Table 1. *Conduction velocities of vagal afferent fibres* (PAINTAL 1953 c and unpublished observations and IGGO 1958)

Type of afferent fibre	No. of fibres	Range (m/sec)	Mean (m/sec)	S.D.
Pulmonary stretch (slowly adapting) . .	57	14—59	36	10
Pulmonary stretch (rapidly adapting) = rapidly adapting tracheal [WIDDICOMBE 1954 (b)].	10	16—37	25	7
Systemic arterial baroreceptor	16	12—53	33	11
Right atrial type A	16	13—27	20	4
Left atrial type A	8	12—19	16	2
Right atrial type B	20	8—29	17	6
Left atrial type B	17	11—26	18	3
Chemoreceptor	3	7—12	10	2
Ventricular pressure receptor	2	14 and 19	—	—
Gastric stretch receptors		0.5—2.3	—	— (IGGO 1958)
Intestinal tension receptors.		0.5—2.3	—	— (IGGO 1958)
Gastric mucosal chemoreceptors . . .		1—5	—	— (IGGO 1958)
Oesophageal receptors	1	8	—	— (IGGO 1958)

will add to the error by increasing the conduction velocity, especially in the case of the faster fibres.

Another source of error is that likely to be introduced by making corrections for the temperature of the vagus when it deviates from 37° C. The Q_{10} for vagus nerves *in situ* was found to be 1.3 [PAINTAL 1953 (c)]. This figure is lower than 1.5 as suggested by GASSER's observations on the phrenic nerve of the dog (GASSER 1928), and 1.6 as indicated by the results of KIRALY and KRNJEVIĆ (1959). This difference is most probably due to the fact that the temperature of the paraffin pool does not represent the average temperature of the vagus because of the presence of its normal blood supply, and its close contact with the common carotid artery through which blood flows at body temperature which is usually maintained at 37° C. Ideally, therefore, it is desirable to keep the pool temperature at 37° C. This is easily accomplished without additional instrumentation in tropical countries.

Other errors relate to the fibre being in a relative refractory state when the electrically evoked impulse is initiated at the stimulating electrodes and errors arising from uncertainty about the start of the impulse. These can be overcome by using suitable precautions [Paintal 1953 (c); Iggo 1958].

Although it might be possible to get a rough estimate about the relative sizes of afferent fibres from the spike heights in thick filaments or in whole nerves (Hunt 1951), spike height can be a misleading criterion in the case of thin filaments containing a few live fibres owing to variations in recording conditions [see Fig. 12 in Paintal 1953 (c)]. In fact it is not uncommon to find, as pointed out by Iggo (1958), that some non-medullated fibres may have larger spike heights than medullated ones in the same strand.

V. Conduction velocities of afferent fibres

The conduction velocities of most of the afferent fibres of the vagus are now known and it is now fairly certain that they range from 0.5—60 m/sec [see Paintal 1953 (c) and Iggo 1958]. Iggo has provided strong evidence to show that the fibres with conduction velocities less than about 2.5 m/sec are C fibres since the properties of their action potentials and stimulus thresholds are quite different from those of medullated fibres. According to Grundfest and Gasser (1938) and Gasser (1950) the maximum conduction velocities of non-medullated fibres is 2.5 m/sec.

Initially [Paintal 1953 (c)] owing to the assumption that the methods of dissection would favour the isolation of thicker fibres a frequency distribution of conduction velocities of all afferent fibres of the vagus was not attempted. However, this bias for thicker fibres is apparently not an important factor because frequency distributions of conduction velocities have been made in muscle nerves and these seem to fit the nerve fibre diameter spectrum (Hunt and Kuffler 1951; Kuffler, Hunt and Quilliam 1951; Hunt 1954). In these plots, the practice has been to divide the conduction velocities determined experimentally by 6, this being the conversion factor from conduction velocity to fibre diameter (Hursh 1939). This procedure seems to work satisfactorily although it is likely to err in the small nerve fibre diameter ranges owing to the conversion factor being smaller (Gasser and Grundfest 1939).

Since the vagal afferent fibres were isolated at random irrespective of the modality of the fibre [Paintal 1953 (c)], it will be in order to plot the frequency distribution of the conduction velocities. This has been done in Fig. 4 in which the nerve fibre diameters plotted from the data of Agostoni et al. (1957) are also given. The number of fibres in each group has been expressed as a percentage of the total number isolated (116). Fig. 4 shows that the electrophysiological data fits the histological data rather well. The close correspondence indicates that large fibres were not isolated preferentially because many more relatively slowly conducting fibres were isolated. If this correspondence is genuine, which there is no reason to doubt, it indicates that the conversion factor of 6 is applicable to vagal afferent fibres and therefore to fibres in the aortic nerve (see p. 79) and that it might be applicable in the lower ranges of conduction velocities as well.

Although the conduction velocities of the myelinated afferent fibres are practically the same in different parts of the cervical vagus (IGGO 1958) they fall in most fibres in the distal part of their course [PAINTAL 1962 (a)]. There are some in which there is no reduction right up to the ending as in the case of muscle stretch receptors [PAINTAL 1959 (a)]. Existing information suggests that although there is some reduction in the main vagal trunk inside the chest, most of the reduction in conduction velocity occurs after the afferent fibres leave the main vagal trunk. Judging from the data from cardiovascular afferent fibres it may be concluded that the reduction in conduction velocity in most fibres will range from 10 to 40% and that there must be some, say about 10%, in which the reduction will be about 50% of the cervical conduction velocity. In any case it would appear that most of the fibres remain myelinated till their terminations [PAINTAL 1962 (a)]. These conclusions apply to afferent fibres with cervical conduction velocities greater than 10 m/sec. There is no information about those with slower con-

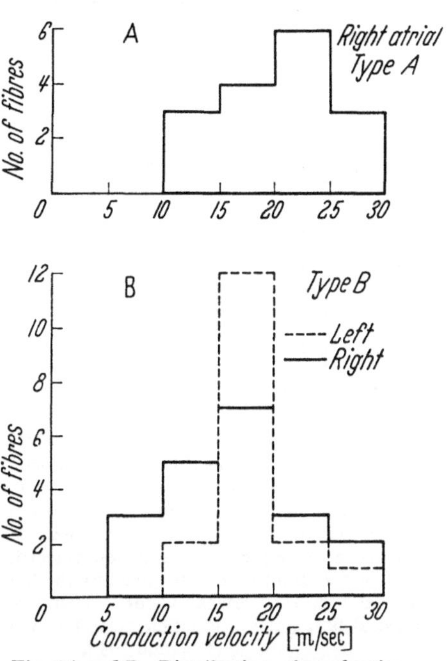

Fig. 6 A and B. Distribution of conduction velocities of afferent fibres from right atrial type A receptors (A) and right and left atrial type B receptors (B). The graphs have been plotted from earlier data [PAINTAL 1953 (c), and certain unpublished observations 1961]

duction velocities. For technical reasons (see PAINTAL 1962) different kinds of experiments designed to determine intrathoracic conduction time will have to be devised for such fibres because there are few cardiovascular mechanoreceptors with fibres conducting so slowly [PAINTAL 1953 (c)].

The conduction velocities of the fibres from various sensory receptors are given in Table 1. A noteworthy feature is that there is a wide range of conduction velocities for most of the fibres, the exception being those from chemoreceptors, deflation receptors and gastro-intestinal receptors. Fibres from pulmonary stretch receptors and systemic arterial baroreceptors have the widest range. The conduction velocities of most of the fibres from pulmonary stretch receptors is above 30 m/sec (Fig. 8B). In contrast no fibre from atrial receptors has a conduction velocity greater than 29 m/sec (Fig. 6).

Table 1 shows that the only fibres with conduction velocities greater than 40 m/sec are those arising from pulmonary stretch receptors and systemic arterial baroreceptors. Presumably there are not many fibres from the latter in the vagus itself. This being so, it follows that if the vagus is separated from the aortic nerve near the nodose ganglion and only the central end of the vagus is stimulated, it should be possible to stimulate exclusively fibres from pulmonary stretch receptors with stimuli that are a little above threshold for pulmonary stretch fibres as done by WYSS (1954). In this connection it would be desirable to monitor the compound action potential of the vagus simultaneously as has been done by WYSS and RIVKINE (1950), WYSS (1954) and PÓRSZÁSZ et al. (1960) and SUCH and PÓRSZÁSZ (1960). Actually with suitable stimulus intensities it should be possible to stimulate more than a third of all fibres from pulmonary stretch receptors before fibres from functionally different endings are also stimulated (see Figs. 3C, 6 and 8B).

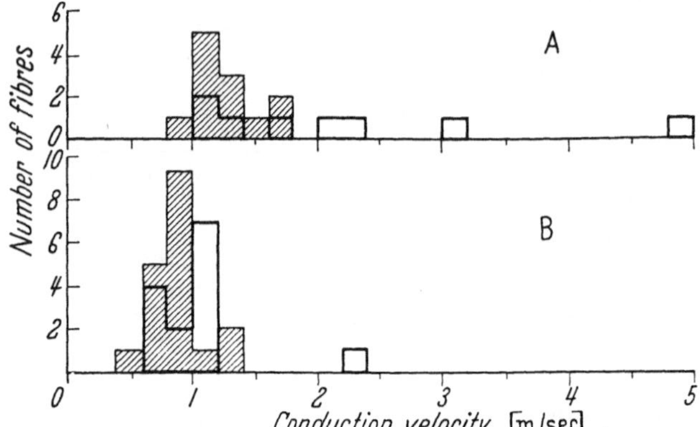

Fig. 7A and B. Conduction velocities in single fibres dissected from the cervical vagus of the cat. A, centripetal fibres from gastric mucosal chemoreceptors; B, centripetal fibres from gastric and intestinal "in series" tension receptors. The cross-hatched areas represent conduction velocities in the abdomen and thorax, and the areas bounded by thick lines the conduction velocities in the cervical vagus. With two exceptions, both mucosal units, the conduction velocities were below 2.5 m/sec (IGGO 1958)

As mentioned already [PAINTAL 1953 (c); WHITTERIDGE 1953] the use of stronger stimuli designed to bring into action the so-called delta or B elevations, is not to be advocated for studies of reflex effects because at such strengths, fibres from 9 different kinds of endings will be activated, i.e. pulmonary stretch receptors, three kinds of tracheobronchial receptors [WIDDICOMBE 1954 (b)], systemic arterial baroreceptors, atrial type A and type B (both from right and left side with different reflex functions), pulmonary arterial and ventricular pressure receptors.

Table 1 also shows that the majority of non-medullated afferent fibres arise from endings in the gastro-intestinal tract. While the conduction velocities of some, e.g. those from gastric stretch receptors, might seem to be exclusively in the C fibre range (Fig. 7B), others such as the oesophageal endings and the gastric mucosal chemoreceptors (Fig. 7A) have fibres in both the medullated and non-medullated ranges of conduction velocities. Recent observations (unpublished) suggest that this might also be true of pulmonary deflation receptors. In view of this it will be reasonable to expect that the lower limit of conduction velocities of the slowest conducting medullated vagal afferent fibres, such as those from chemoreceptors, might also run into the C fibre range. In this connection the results of DOUGLAS, RITCHIE and SCHAUMANN (1956) are of interest.

VI. Cold block of vagal afferent fibres

Low temperatures have been used as a means of blocking conduction in afferent fibres so that some reflexes can be eliminated and others studied in relative isolation [e.g. HEAD 1889; WHITTERIDGE and BÜLBRING 1944; WIDDICOMBE 1954 (c)]. This procedure has yielded valuable information,

Table 2. *Range of conduction velocities of vagal afferent fibres and the range of vagal tempe-ratures required to block conduction in them*

Type of afferent fibre	Range of blocking temperature (^{0} C)	Range of conduction velocities (from Table 1) (m/sec)
Pulmonary stretch	8—19 * +	14—59
Slowly adapting bronchial . .	7—13 *	
Rapidly adapting tracheal . .	6—16 *	16—37
Intermediate tracheobronchial	7—10 (3 fibres) *	?
Systemic arterial baroreceptor	8—12 + x	12—53
Type A atrial	8—12 + x	12—27
Type B atrial	4—8 + x	8—29

x = TORRANCE and WHITTERIDGE (1948).
+ = WHITTERIDGE (1948).
* = WIDDICOMBE [1954 (b)].

such as the existence of HEAD's paradoxical reflex (HEAD 1889) and the presence of reflexes evoked by the injection of various chemical substances (see DAWES and COMROE 1954). However, difficulties arise when attempts are made to correlate the disappearance of a particular reflex with block of conduction in afferent fibres of a particular size, because it is usually assumed that cooling the vagi might be expected to block the larger fibres first (DAWES, MOTT and WIDDICOMBE 1951). This assumption is justified if differential block of A and C fibres is involved, because LUNDBERG (1948) and DOUGLAS and MALCOLM (1955) have shown clearly that conduction in C fibres survives after it has been blocked in A fibres.

The difficulty only arises with regard to differential blocking of the medullated fibres themselves. Here direct evidence based on observations of the compound action potential shows that in the A group, the delta fibres are the first to be blocked and that the larger medullated fibres are blocked at lower temperatures (DODT 1953; DOUGLAS and MALCOLM 1955). In the saphenous nerve the order of block is: delta, gamma, beta and finally alpha (DOUGLAS and MALCOLM 1955). However, in the case of the vagus, DOUGLAS and MALCOLM (1955) could not arrive at a firm conclusion because there was no similar clear-cut differential effect. This was explained as being due to there being a wide range of conduction velocities contributing to the A and B components of the compound action potential which made assessment of the results difficult by the methods used by them.

On the other hand there is direct evidence obtained by TORRANCE and WHITTERIDGE (1948), WHITTERIDGE (1948), WIDDICOMBE [1954 (b)] and HENRY and PEARCE (1956) about the blocking temperatures of afferent fibres from different endings. These are given in Table 2 which also shows the range of conduction velocities of the fibres. Table 2 shows that apart from some fibres from type B atrial receptors, conduction in various fibres does not survive cooling below 7—8⁰ C. It is perhaps also significant that all fibres from arterial baroreceptors, and atrial type A and type B receptors conduct if the temperature of the nerve is above 12⁰ C. On the other hand it is clear from Fig. 5 of WIDDICOMBE's paper [1954 (b)] that 84% of all fibres from pulmonary stretch receptors are blocked at temperatures above 12⁰ C (12 to 19⁰ C). Cold block at 12⁰ C therefore forms an important dividing line above which most fibres of atrial receptors, rapidly adapting tracheobronchial receptors [see Fig. 5 in WIDDICOMBE 1954 (b)], and arterial baroreceptors will conduct, but most of those from pulmonary stretch receptors will be blocked.

One can now consider the relation of blocking temperatures to the conduction velocities of the fibres. There are three possibilities. Firstly, that as in the case of the saphenous and lingual nerves, the slowest medullated fibres will be blocked first followed by the faster conducting fibres, and last of all the fastest A fibres (DODT 1953; DOUGLAS and MALCOLM 1955). The second possibility is the converse of the first, namely, that the fastest fibres are blocked first and last of all the slowest ones. Thirdly it is possible that there is no relation between conduction velocities and the temperatures at which block occurs in medullated afferent fibres of the vagus nerve.

Table 2 shows that the lower limit of conduction velocities of all the afferent fibres mentioned above is about 8—16 m/sec, i.e. it may be regarded as being nearly the same for all fibres within this range. It therefore follows, that the upper limit of cold block should be nearly the same in all these fibres if the first possibility is correct. Table 2 shows that this is not so. Conversely, according to the first possibility one would expect that the lower limit of cold block of pulmonary stretch fibres will be decidedy lower than that for the other fibres especially those from atrial receptors because the conduction velocities of pulmonary stretch fibres are higher. In fact there should be an obvious difference in the lower limit of cold block for pulmonary stretch fibres on the one hand, and all the other fibres on the other because whereas 72% of all fibres from pulmonary stretch receptors have conduction velocities greater than 30 m/sec [30—59 m/sec (see Fig. 8B)], there are no fibres from atrial receptors with conduction velocities greater than 29 m/sec (Fig. 6); only a fifth of those from rapidly adapting tracheobronchial receptors (see Fig. 8B) and half of those from systemic arterial pressure receptors (Fig. 3 C) have conduction velocities greater than 30 m/sec. These results lead one to expect that conduction in fibres from pulmonary stretch receptors should be blocked at much lower temperatures than in the other fibres. But this is not the case (Table 2) and so the first possibility is invalidated.

On the other hand the results can be explained quite satisfactorily on the basis of the second possibility, namely that the faster conducting fibres are blocked first and finally the smallest medullated fibres. In fact it would appear that 12⁰ C will block most fibres conducting faster than 30 m/sec. This will explain why the majority of pulmonary stretch fibres are blocked by temperatures between 18 and 12⁰ C and why fibres from atrial receptors continue to conduct at 12⁰ C. Additional support comes from the observations concerning the differential blocking temperatures of fibres from pulmonary stretch receptors and rapidly adapting tracheobronchial receptors with an adaptation index of 100 % (Fig. 8).

The blocking temperatures of these fibres are shown in Fig. 8 A which has been plotted from the data contained in Fig. 5 of WIDDICOMBE's paper [1954 (b)]. In Fig. 8B are shown the distribution of conduction velocities of the fibres concerned [PAINTAL 1953 (c)]. The similarity of the two sets of curves is quite remarkable which leads one to conclude that the blocking temperatures of the fibres are related to their conduction velocities, the greater the conduction velocity, the greater the blocking temperature. If this is true one could speculate from Fig. 8 that the blocking temperature might even be linearly related to the conduction velocities of the fibres, and that between conduction velocities of 10 and 60 m/sec, an increase in the conduction velocity of the fibre by about 4.5 m/sec would increase the blocking

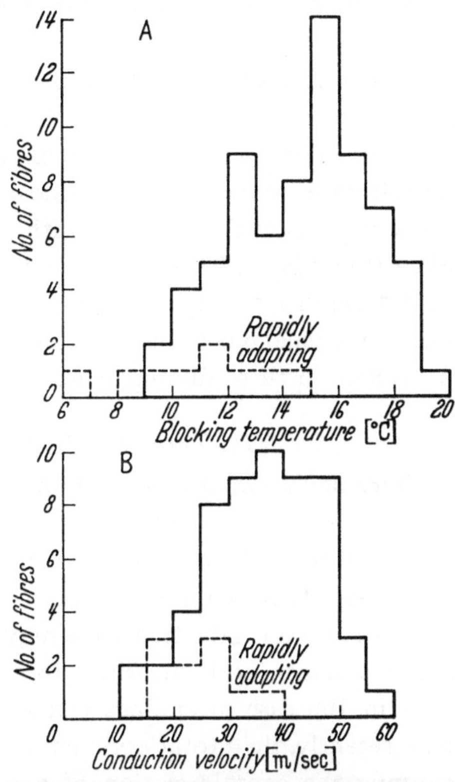

Fig. 8 A and B. Blocking temperatures, A and conduction velocities B, of afferent fibres from pulmonary stretch receptors and rapidly adapting tracheal receptors. The data in A for vagal temperature needed to block conduction in individual afferent fibres was obtained from Fig. 5 in WIDDICOMBE [1954 (b)]. In A and B only those rapidly adapting fibres are shown which have an adaptation index of 100 %. The conduction velocities are as originally reported [PAINTAL 1953 (c)]

temperature of the fibre by about 1⁰ C. These conclusions need to be confirmed experimentally.

However, there are certain discrepancies that need to be borne in mind. Firstly, it is surprising that Torrance and Whitteridge (1948) and Whitteridge (1948) did not encounter fibres from systemic arterial baroreceptors that were blocked at temperatures higher than 12⁰ C since about half the total number of fibres have conduction velocities greater than 30 m/sec. Secondly, the difference between the blocking temperatures of type A and type B receptors (Table 2) is difficult to explain since the conduction velocities of the fibres from these receptors are almost identical (Fig. 6) although there are some fibres from type B receptors with conduction velocities in the 5—10 m/sec range (Fig. 6B). The difference could also be attributed to uncertain experimental conditions such as thermode calibration [cf. Widdicombe 1954 (b)] which may yield incongruous results. In any case more experiments are needed to explain all these discrepancies.

VII. Total conduction time in afferent fibres

Since nerve impulses are commonly recorded somewhere along the cervical vagus and the various endings are located some distance away, it is important to get an estimate of the total conduction time from the endings to the recording electrodes. This factor is of special importance in non-medullated fibres in which the total conduction time may be as large as 0.2—0.5 sec from endings in the stomach and intestines to the recording electrodes [Iggo 1957 (a)]. Even in the case of endings connected to medullated fibres, it has become clear recently that total conduction time must be taken into account when studying the causal relation of a specific natural stimulus to a particular discharge of impulses (Paintal 1962).

Total conduction time can be determined by several methods. If the nerve fibre can be stimulated somewhere close to the ending it can be determined by stimulating the fibre electrically. However, this method can be quite erroneous (see Paintal 1962) except in the case of non-medullated fibres in which the total conduction time is relatively very large. Besides, it is necessary to open the chest or abdomen in order to stimulate the fibre locally, and this itself may cause an increase in total conduction time. Further, by local stimulation it is not possible to determine the conduction time between the stimulating electrodes and the ending.

Only indirect methods are available for determining total conduction time precisely [Paintal 1959 (a), 1962]. One of these methods requires that the ending should be amenable to mechanical stimulation by short-lasting pulses and that it should also have a steady discharge of impulses [Paintal 1959 (a)]. A less precise method has to be used in the case of cardiovascular endings because they cannot be stimulated consistently with short pulses and they cannot be made to discharge a continuous series of impulses. The latter method requires that successive bursts of impulses in cardiovascular afferent fibres should be similar over at least 4—6 cardiac cycles (Fig. 9). In addition it is necessary to use special methods to take into account fluctuations in the interval between impulses (Fig. 9). The methods which consist of determining graphically (Fig. 9E) the time taken for an antidromic impulse to travel from the stimulating electrodes placed in the distal part of the neck, to the ending often involves several hundred measurements of time intervals. The method is accurate to within 0.5 msec.

The intrathoracic conduction time computed in this way corresponds to the minimum predicted intrathoracic conduction time in about half of the cardiovascular afferent fibres. In about a third it exceeds the minimum predicted value by less than 1.3 msec and rarely it may exceed it by several milliseconds. This means that in the majority of afferent fibres of the heart and lungs the intrathoracic conduction time will not exceed the minimum predicted value by about 1 msec. Thus if the cervical conduction velocity of an afferent fibre is known and its location is also known it is possible to predict with some confidence what the total conduction time from the ending to the recording electrodes will be (PAINTAL 1962). If the conduction velocity is not known, but the impulses are recorded near the nodose ganglion, it is possible to get a rough estimate of what the total conduction time is likely to be from the existing information available about the cervical conduction velocities of various afferent fibres (Table 1). Thus in the case of atrial receptors, the total conduction time will range from 4—19 msec and in about 70% of the fibres it will be less than 10 msec (PAINTAL 1962). In the case of systemic arterial baroreceptors many of which are located near the brachiocephalic trunk, the total conduction time from the ending to the nodose ganglion will range from 1—5 msec and in most fibres it will be less than 2 msec. Since pulmonary stretch receptors appear to be located in the bronchi and bronchioles [WIDDICOMBE 1954 (d)], the intrathoracic conduction distance will be roughly comparable to that in atrial afferent fibres, so that the total conduction time should range from 2—12 msec and in most fibres it will be probably less than 6 msec.

All the above figures relate to measurements with the closed chest. With open chests the intervals will be longer owing to the reduction in conduction velocity in the intrathoracic part of the afferent fibres [PAINTAL 1962]. The values will be essentially the same if the impulses are recorded in the vagal rootlets because the conduction distance between the nodose ganglion and vagal rootlets is small compared to the total conduction distance from the endings to the nodose ganglion. This means that the temporal relations of cardiovascular discharges to the ECG will be the same in the vagal rootlets as it is near the nodose ganglion so that existing criteria for identification

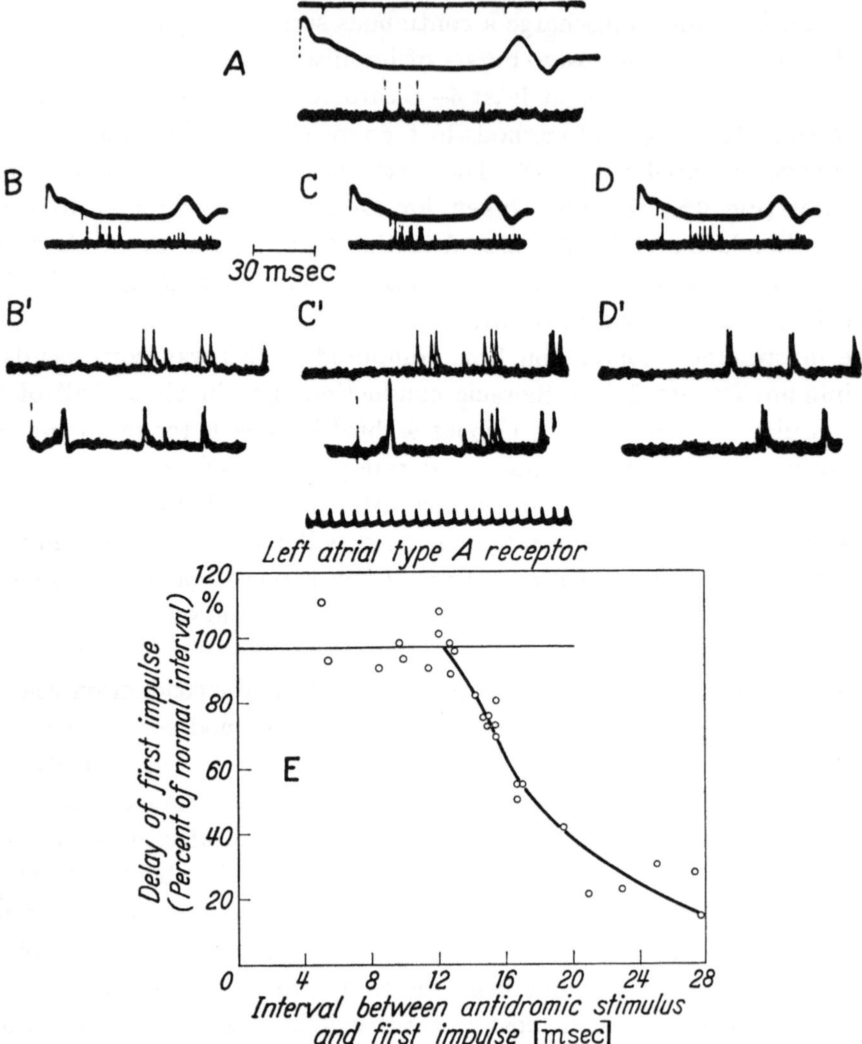

Fig. 9A—E. Left atrial type A receptor. Records B to D show the use of alternate sweep stimulation (see Paintal 1962) for determining intrathoracic conduction time in the fibre. Sweeps were triggered by the large P wave of intra-atrial lead 2. A, shows the normal pattern of discharge of 3 impulses in a burst. B, shows 4 superimposed sweeps recorded on one c.r.t. and B′ 2 superimposed sweeps (expanded portions of sweeps in B) on each of two traces on another c.r.t. The sweeps with an antidromic stimulus are displaced downwards. In B′ the antidromic stimulus was subthreshold for the fibre in order to show that the distribution of impulses in the two sets of sweeps is symmetrical. In C and C′ the stimulus was suprathreshold so that the antidromic impulse collided with the first impulse (note its absence in the lower sweeps of C′) and left the moment of appearance of the second impulse unaffected as shown by the identical average moment of appearance of the second impulse in the upper and lower sweeps of C′ with respect to the small electrical artifact. In D and D′ the antidromic impulse arrived before the initiation of the first impulse thus delaying the appearance of the first re-set impulse in the lower trace in D′. Millisecond time marks at the bottom belong to the expanded sweeps. E is a graph showing the delaying effect of an antidromic impulse on the initiation of the first impulse in the same receptor shown in B to D′. The abscissa represents the interval between the antidromic stimulus and the normal time of appearance of the first impulse obtained from sweeps such as those shown in B′, C′ and D′. The ordinate represents the delay in the appearance of the first impulse expressed as a percentage of the normal interval between the 1st and 2nd impulses. The graph begins to slope downward at 12.0 msec which represents the interval between the antidromic *stimulus* and the normal moment of appearance of the first impulse when the antidromic *impulse* has an even chance of colliding with the first impulse or arriving at the ending just before it is initiated. This interval is = $2 t_i + t_0$. Minimum predicted value of this was 10.9 msec. Computed t_i (intrathoracic conduction time) is 4.7 msec (t_0 = conduction time in cervical vagus) (Paintal 1962)

of different fibres are applicable if the impulses are recorded in the rootlets or even in the medulla (PAINTAL 1952; BONVALLET and SIGG 1958; BAUMGARTEN and ARANDA 1959; BAUMGARTEN, KOEPCHEN and ARANDA 1959; HELLNER and BAUMGARTEN 1961).

Compared to the above figures, the total conduction time in gastric and intestinal fibres is very large because the conduction velocities of most of them are less than 1.5 m/sec (IGGO 1958) and the conduction distance from the stomach and intestines is of the order of 30 cm. The total conduction time will therefore be greater than 150 msec in most fibres and will be about 0.5 sec in the slowest ones. This is what IGGO has observed [IGGO 1957 (a)].

VIII. Functional subdivisions of vagal afferent fibres

Luckily, in the case of the vagus the practice of subdividing nerve fibres into groups as in the case of muscle nerves (see LLOYD 1943; GRANIT 1955) or skin nerves (ERLANGER 1928) has not come into vogue, so that the confusion that now exists in the case of afferent fibres of muscle and skin nerves [see Table on p. 260 of WRIGHT's text book of Physiology (KEELE and NEIL 1961)] with overlapping of conduction velocities of fibres from different sense organs is not present in the vagus even though this nerve contains afferent fibres from at least 16 different kinds of endings as against only 5 known kinds in muscle nerves. This confusion has been avoided because since ADRIAN's paper on the vagus (ADRIAN 1933) the practice has been to designate afferent fibres according to the type of endings they come from. Thus it is the practice to speak in terms of pulmonary stretch fibres, deflation fibres, etc. Accordingly, the afferent fibres of the vagus may be divided into the following functional groups:

A. Pulmonary afferent fibres from:
1. Slowly adapting pulmonary stretch receptors (ADRIAN 1933).
2. Slowly adapting bronchial stretch receptors [WIDDICOMBE 1954 (b)].
3. Rapidly adapting tracheal receptors [WIDDICOMBE 1954 (b)]. These are identical with KNOWLTON and LARRABEE's rapidly adapting receptors (KNOWLTON and LARRABEE 1946).
4. Intermediate tracheobronchial receptors [WIDDICOMBE 1954 (b)].
5. Specific deflation receptors [PAINTAL 1955 (a)].

B. Mediastinal afferent fibres [ADRIAN 1933; WIDDICOMBE 1954 (b)].

C. Cardiovascular afferent fibres from (see HEYMANS and NEIL 1958):
1. Systemic arterial baroreceptors.
2. Chemoreceptors.
3. Type A atrial receptors.
4. Type B atrial receptors.
5. Ventricular pressure receptors.
6. Pulmonary arterial baroreceptors.

D. Gastro-intestinal and oesophageal afferent fibres from:
   1. Gastric stretch receptors [Paintal 1953 (d), 1954 (b)] which are referred
      to as tension receptors by Iggo [Iggo 1955, 1957 (a)].
   2. Gastric mucosal chemoreceptors [Iggo 1957 (b)].
   3. Intestinal tension receptors [Iggo 1957 (a)].
   4. Intestinal mucosal mechanoreceptors [Paintal 1957 (c)].
   5. Oesophageal tension receptors [Iggo 1957 (a); Andrew 1957].

As will become clear in the following pages the afferent fibres terminating
in a particular group of sensory endings have a characteristic pattern of
discharge which is aroused by specific stimuli and which distinguishes them
from other afferent fibres although the conduction velocities of various fibres
may overlap considerably. These will now be described one by one.

IX. Pulmonary afferent fibres

While recording impulses from the cut peripheral end of the whole vagus,
the impulses from pulmonary stretch receptors overshadow the activity of
all other fibres, and when the discharge is heard over a loud speaker as is
conventional in studies of this sort, a respiratory rhythm, increasing during
inspiration, can be heard clearly. Such activity was described by Adrian,
in 1926 and by Keller and Loeser (1926) and later by Partridge (1933)
but it was not till 1933 when Adrian described unitary discharges that a
clearer understanding of the mode of activation of the endings was obtained.
Adrian also described some discharges elicited by deflation of the lungs.
His description covered many of the essential aspects, so that there seemed
little need for further work till Knowlton and Larrabee (1946) differentiated
responses from pulmonary endings into rapidly and slowly adapting types
(Fig. 10) with presumably different reflex effects (Larrabee and Knowlton
1946). This led to some speculation about their mode of excitation since the
rapidly adapting ones were also stimulated by deflation and the situation
remained somewhat confused till Widdicombe [1954 (b)] finally showed
that the rapidly adapting receptors of Knowlton and Larrabee (1946) were
tracheobronchial endings, and that some of Adrian's deflation receptors were
most probably slowly adapting tracheobronchial receptors. Finally, the
existence of specific deflation receptors was revealed by the use of chemical
substances [Paintal 1955 (a)]. Currently therefore, the 5 types of pulmonary
endings enumerated above have come to be recognised.

A. Pulmonary stretch receptors

The first detailed description of the discharges from pulmonary stretch
receptors was given by Adrian (1933). These endings are stimulated only
during normal inspiration and by inflation of the lungs (Fig. 10A). Their

afferent fibres are the easiest to isolate as "single units" and they are easily identified by noting the increase in activity during normal inspiration. Difficulty in identification may arise only if the endings have a superimposed cardiac rhythm because it is then necessary to distinguish the endings from cardiovascular receptors which have a respiratory rhythm as well, notably

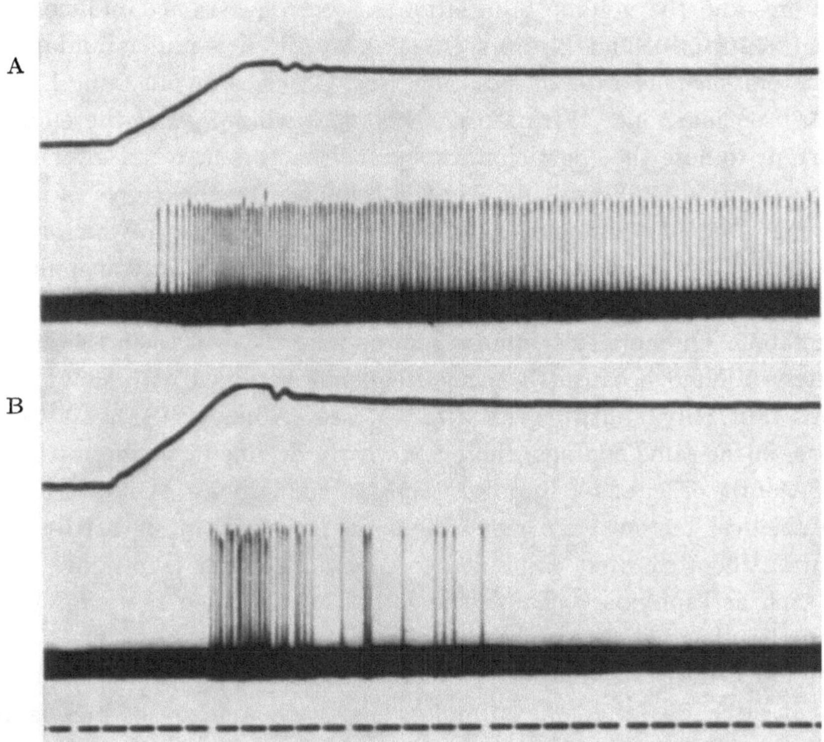

Fig. 10 A and B. Responses of two kinds of afferent fibres in the vagus to inflation of the lungs. The first receptor adapts slowly, the second very rapidly to maintained inflation of the lungs. Chest wall removed. Respiration pump stopped in expiration just before start of each record. The upper line represents intratracheal pressure (increased pressure upwards) and the lower line marks time in $^1/_2$ and $^1/_{10}$ seconds (KNOWLTON and LARRABEE 1946)

the type B atrial receptors [WHITTERIDGE 1948; PAINTAL 1953 (a)]. However, most often, it is easy to show that the endings lie in the lungs because they are markedly stimulated by inflation which may knock out the cardiac rhythm [WHITTERIDGE 1948; BIANCONI and GREEN 1959 (c)].

For details concerning the responses of the endings, ADRIAN's paper should be consulted (ADRIAN 1933). In the frog impulses from pulmonary stretch receptors have been recorded by NEIL, STRÖM and ZOTTERMAN (1950), BONHOEFFER and KOLATAT (1958) and DOWNING and TORRANCE (1961).

Natural stimulus of the endings. ADRIAN showed conclusively that the discharge from pulmonary stretch receptors is linearly related to the volume of inflation of the lungs. This has been confirmed repeatedly, e.g. Whitte-ridge and Bülbring (1944) and Widdicombe [1954 (d)] and also by Bon-hoeffer and Kolatat (1958) in the frog. The relationship is particularly noteworthy in the frog because unlike the cat, intratracheal pressure and volume are not linearly related under static conditions (see Fig. 5 in Bon-hoeffer and Kolatat 1958). However, this relationship between activity in the fibres and the volume of inflation has been questioned on inconclusive evidence (Weidmann and Bucher 1951) and while it is understandable that other factors such as rate of inflation (see Davis, Fowler and Lambert 1956), or bronchial tone [Widdicombe 1954 (d)] will influence the endings, it is important to note that participation by such factors does not alter the fact that the activity of endings is determined primarily by the degree of inflation of the lungs. As pointed out by Adrian (1933), the manner in which inflation excites the endings is not known. It could be by increase in transpulmonary pressure as suggested by Davis et al. (1956).

Adaptation. Pulmonary stretch receptors adapt slowly when the lungs are moderately inflated because the peak frequency attained with slow or rapid inflations falls only a little even after 10 sec (Adrian 1933). With large inflations, in the same endings, there is an early decline in the high frequency discharge (of the order of 300 impulses/sec) attained initially. Adrian suspected that mechanical factors were responsible for this, such as redistribution of air allowing the areas first expanded to contract. Other factors are also involved such as asphyxia, because the initial fall is much more rapid when the circulation has failed (Adrian 1933).

In practice the term adaptation merely indicates the decline in the frequency of discharge after the stimulus has reached its peak and is being maintained at that level. This may be a property of the ending itself which can be varied by chemical means [Whitteridge and Bülbring 1944; Gray and Diamond 1957; Paintal 1957 (b)], or part of it may be due to mechanical factors, e.g. in muscle spindles (Matthews 1933), cray fish stretch receptors (Krnjević and van Gelder 1961), etc. In the case of pulmonary stretch receptors mechanical factors dependent on transpulmonary pressures and air flow are involved (Davis et al. 1956). To this may be added the effect of bronchial tone [Widdicombe 1954 (d)]. Conversely, when the stimulus is withdrawn, i.e. when the lungs deflate, there is an overshoot in the reduction of the discharge. The more important factor responsible for this overshoot, is probably related to the ending itself contributing to the dynamic "off effect" (Katz 1950) which is clearly demonstrable in the frog's muscle spindle. This factor which largely determines the end of activity in aortic and carotid baroreceptors [Landgren 1952 (a), (b)] and which Davis et al. (1956) did not take fully into consideration, must play an important part in determining the cessation of the discharge in pulmonary stretch receptors.

Role of other factors. Although the threshold of these endings can sometimes be changed by altering the position of the lungs (ADRIAN 1933), this is not an important factor that needs consideration except perhaps when differentiating the endings from mediastinal receptors [WIDDICOMBE 1954 (b)]. On the other hand changes in bronchial tone can influence the response of the endings markedly and consistently [WIDDICOMBE 1954 (d)]. Thus bronchoconstriction increases the discharge and bronchodilatation reduces it. The excitatory effects of 5-HT in cats and dogs but not in rabbits (SCHNEIDER and YONKMAN 1953, 1954) can probably also be explained in the same way. Although so far these effects have been produced by drugs injected intravenously, it is presumed that similar effects will occur when changes in bronchial tone are produced reflexly (WIDDICOMBE 1961) or in pathological states such as asthma.

WIDDICOMBE [1954 (d)] suggests that the effects of the drugs are exerted on the smooth muscle in which the endings lie and correlates this with histological evidence (LARSELL 1921; ELFTMANN 1943). He believes that the slowly adapting endings might be smooth muscle spindles and that the increase in discharge during bronchoconstriction is not due to the *local* contraction of the muscle but to the increase in flow resistance giving rise to higher intratracheal pressures. As indicated by WIDDICOMBE [1954 (d)] and stressed by DAVIS at al. (1956), it appears that the frequency of discharge depends on transpulmonary pressures.

Effect of pulmonary congestion. CHRISTIE's suggestion that cardiac dyspnoea accompanying pulmonary congestion might be due to sensitization of pulmonary stretch receptors owing to increased stiffness of the lungs seemed so plausible (CHRISTIE 1938), that BÜLBRING and WHITTERIDGE (1945) were surprised to find very small increases in activity of pulmonary stretch receptors during pulmonary congestion produced by impeding venous outflow or by raising the pulmonary arterial pressure in perfused lungs of the cat. The problem was reinvestigated by MARSHALL and WIDDICOMBE (1958) in cats with closed chests and they found that there was an increase of about 20% in the frequency of discharge even when an allowance was made for the increase in the functional residual capacity. Similar increases were observed by COSTANTIN (1959). This increase was attributed to increased stretching of the air passages owing to the reduced compliance, which in the cat is decreased by less than 10% under comparable conditions (HUGHES, MAY and WIDDICOMBE 1958). This implies an increase in intratracheal pressure which MARSHALL and WIDDICOMBE probably did not find. Perhaps unequal ventilation resulting in redistribution of air might play a part as suggested by BÜLBRING and WHITTERIDGE (1945) and MARSHALL and WIDDICOMBE (1958), but this is not likely because an increase in activity was seen in nearly all endings examined by MARSHALL and WIDDICOMBE (1958) and COSTANTIN (1959).

7*

If the small increase in discharge during congestion is explained by changes in the mechanical properties of the lungs, then it follows that the endings themselves are not sensitized chemically by congestion because sensitization means that the endings should yield a higher frequency discharge to the same stimulus, e.g. after trichlorethylene [Whitteridge and Bülbring 1944; Paintal 1957 (b)].

Location of pulmonary stretch receptors. It seems likely that many of the endings lie in the intrapulmonary bronchi [Widdicombe 1954 (d)]. They are not located in the visceral pleura as suggested by Wei mann, Berde and Bucher (1949) because Widdicombe found that activity in the endings survived stripping large areas of the visceral pleura. The main evidence that favours bronchial location of the endings is the effect of increase in bronchial tone by acetylcholine. This drug which constricts the bronchioles, *reduces* the discharge if it is injected *during* a maintained inflation of the lungs. This is believed to be due to the ending being unloaded owing to local contraction of the smooth muscle. However, on closer analysis it seems that Widdicombe's results on the effects of drugs during maintained inflation are equivocal because his Table 2 [Widdicombe 1954 (d)] shows that other bronchoconstrictors such as histamine increased the discharge in both the cases in which it was tried, and eserine also tended to increase the discharge or it produced no change, except in two instances when it reduced the discharge slightly. The effects of adrenaline are also difficult to reconcile with Widdicombe's explanation for the effects of acetylcholine because in 6 out of 7 cases, adrenaline also reduced the discharge instead of increasing it.

The effects of congestion of the lungs on pulmonary stretch receptors also favours bronchial location (see above, Marshall and Widdicombe 1958); and the results of Davis et al. (1956) are best explained if the endings are assumed to lie in the intrapulmonary bronchi and bronchioles.

B. Slowly adapting bronchial receptors

Most of these endings are located in the main bronchus and their responses to inflation closely resemble those of pulmonary stretch receptors, the only distinguishing feature presumably being that they are stimulated by deflation as well (Fig. 11). However, one wonders in what way their responses will differ when impulses are recorded with patent bronchi in which air flow is allowed to occur. Conversely, will the pulmonary stretch receptors be stimulated by deflation if the bronchioles distal to their location are blocked? In practice therefore, without the main bronchus occluded, difficulty will arise in identifying these endings from pulmonary stretch receptors some of which are also excited by deflation.

The threshold of these receptors to inflation [see Fig. 20 in Widdicombe 1954 (b)] indicates that more than a third of the endings will be active during the respiratory pause because the intrathoracic bronchi are subjected to intrapleural pressures (Fig. 11 A, B). During inspiration these endings will therefore respond like typical pulmonary stretch receptors and will in fact be signalling the amount of air in the lungs. In view of this it may not be justifiable to regard them as a separate group on the basis of their location in the extra-pulmonary bronchi and possibly greater stimulation by deflation, especially

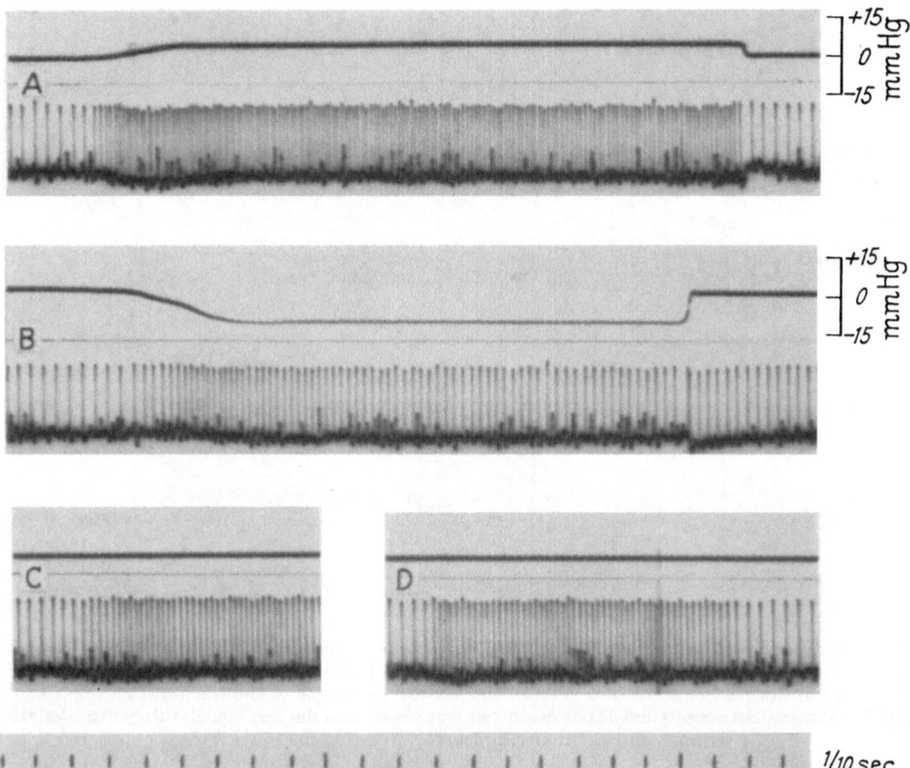

Fig. 11 A—D. A slowly adapting bronchial receptor of a cat. A, distension of the air passages caused an increase in discharge rate; B, deflation caused a similar but smaller effect; C, stimulation of the receptor by touching the outside of the bronchial wall; D, stimulation of the receptor with an endobronchial catheter
[WIDDICOMBE 1954 (b)]

since their reflex effects are similar to those of pulmonary stretch receptors [WIDDICOMBE 1954 (a), (b)]. A possible difference of these endings from pulmonary stretch receptors is that they are only weakly sensitized by ether, but the results are not conclusive as the concentrations which WIDDICOMBE used were much higher than those which sensitize pulmonary stretch receptors (see WHITTERIDGE and BÜLBRING 1944). Consequently since these endings resemble the pulmonary stretch receptors closely, one might lump the two together as broncho-pulmonary stretch receptors.

C. Rapidly adapting tracheal receptors

In contrast, the rapidly adapting endings, which are in fact the rapidly adapting receptors of KNOWLTON and LARRABEE (1946), are clearly different from pulmonary stretch receptors (Fig. 10B). They are not active in eupnoea. They give "on" and "off" responses to both inflation and deflation (Fig. 12A, B) and their threshold to inflation is on the whole much higher than that of other

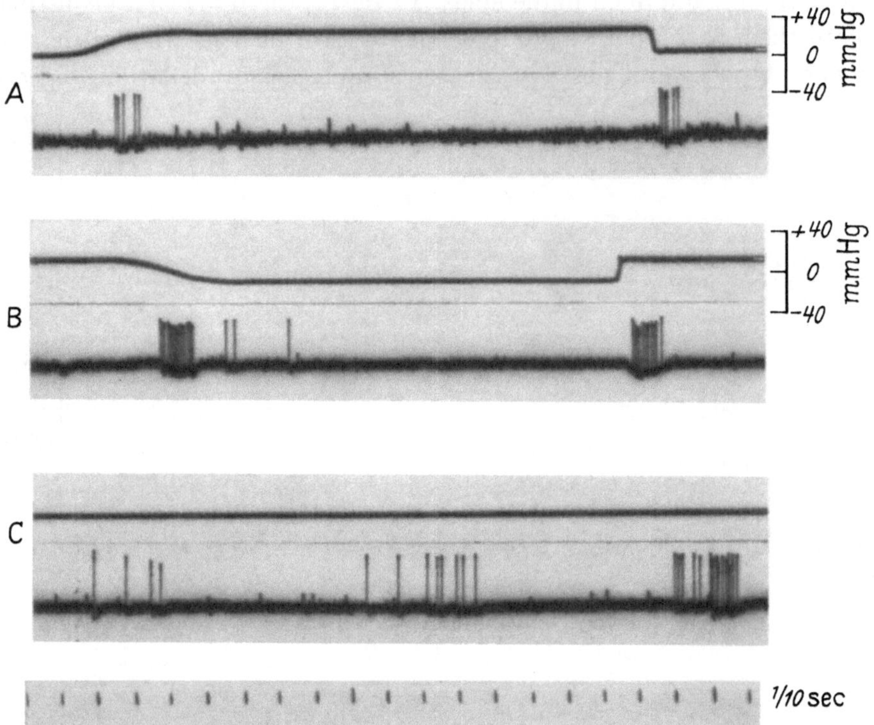

Fig. 12A—E. Rapidly adapting tracheal receptors. A, inflation of the trachea; B, deflation; C, stimulation of the receptor by touch with an endotracheal catheter. D and E show the effect of inhaled talc on another rapidly adapting tracheal receptor of a cat. D, two control deflations of the trachea, showing a typical "double" response. Between D and E powdered talc was blown into the trachea. E, subsequent deflations showing sensitization of the receptor [Widdicombe 1954 (b)]

endings. A characteristic feature of these endings is that they are sensitized by powdered talc or starch (Fig. 12E). Since most of these endings are located in the trachea it is easy to stimulate them mechanically with an endotracheal catheter (Fig. 12C). They are clearly not stimulated merely by air flow unless there is some secretion in the trachea [Widdicombe 1954 (b)]. With the aid of these responses identification of these endings should be easy.

Knowlton and Larrabee (1946) found that about 16% of all receptors which could be stimulated by rapid inflation of the lungs had an adaptation index of 100%. This was confirmed later because it was found that 10 out of 67 fibres examined (see Table 1), i.e. 15%, had an adaptation index of 100% [Paintal 1953 (c)], but Widdicombe found that such fibres constituted only 7% of his population [Widdicombe 1954 (b)]. He ascribes this discrepancy between his results and those of Knowlton and Larrabee to the possibility that Knowlton and Larrabee were not using sufficiently suprathreshold

stimuli because when large inflations are used (up to 250 ml), several endings
which cease to fire before the 2nd second of inflation, can be made to fire
beyond it [WIDDICOMBE 1954 (b)]. For convenience, WIDDICOMBE has

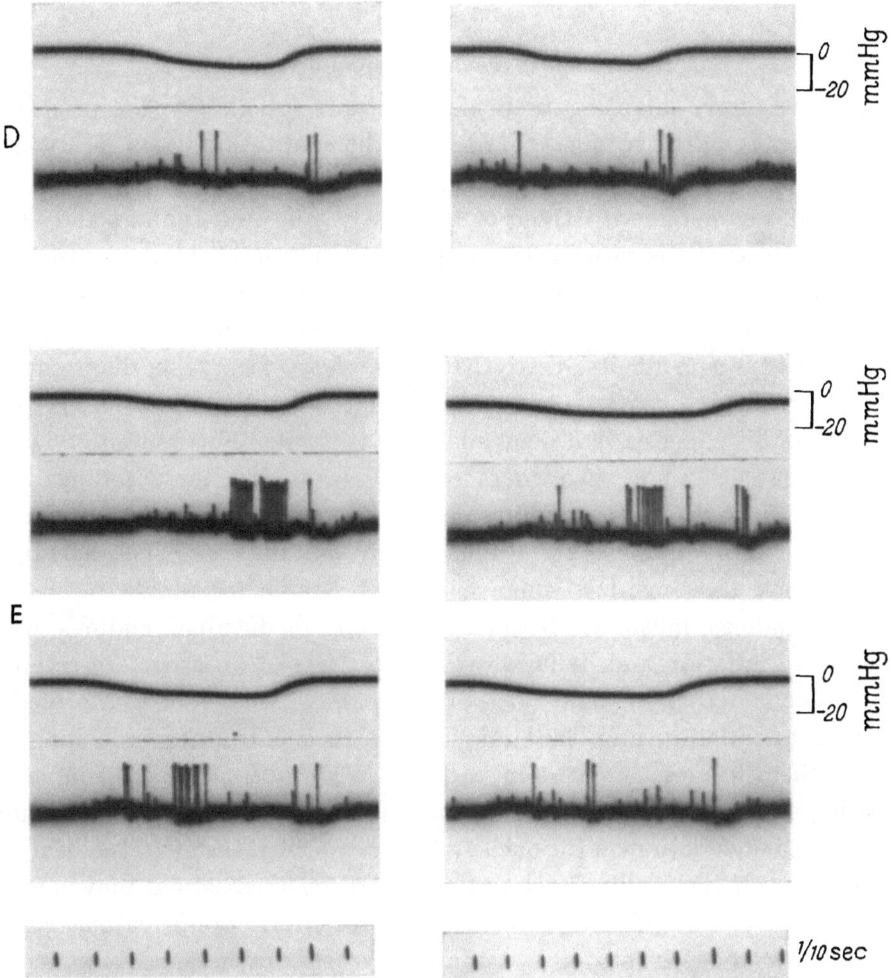

Fig. 12 D and E

classified all endings with an adaptation index of more than 70 as rapidly
adapting.

Natural stimulus. Although these endings adapt rapidly to inflation, the
term rapidly adapting receptors used to describe them is not appropriate
because inflation and deflation are not their natural stimuli except perhaps
during coughing. The natural stimulus of these endings which are *normally
silent*, appears to be local mechanical irritation which may be brought about
by mucus, or dust or other particles (Fig. 12 E).

Conduction velocities. The conduction velocities of fibres from pulmonary
stretch receptors range from 14—59 m/sec (mean = 36 m/sec). Those from

rapidly adapting tracheal receptors range from 16—37 m/sec (mean = 25 m/sec). The lower limit of conduction velocity of the two is therefore similar, but at the upper range there are several fibres from pulmonary stretch receptors with higher conduction velocities (Fig. 8 B).

D. Intermediate tracheobronchial receptors

Like the slowly adapting bronchial receptors (cf. above), the position of these endings is a little confusing if one uses the adaptation index (Knowlton and Larrabee 1946) as a criterion for differentiation, because at one end their responses merge with those of the slowly adapting endings and at the other with those of the rapidly adapting ones. In practice their identification must therefore offer a problem if this is done largely on the basis of adaptation rates to inflation and deflation, especially if the identification is desired during the experiment itself so that some help from the diagrammatic representations of the pattern of response as shown by Widdicombe [1954 (b)] in Fig. 18 of his paper cannot be available at the time. Fortunately, however, Widdicombe has found two characteristic features of these endings. One is that the response of the endings decline on repeated stimulation, and the other is that most of them are stimulated by irritant gases such as sulphur dioxide which also sensitizes some pulmonary stretch receptors.

These endings must be distinguished from mediastinal endings whose responses to inflation and deflation resemble those of intermediate receptors quite closely. Actually, about 80 % of the endings which show intermediate responses to inflation and deflation lie outside the trancheobronchial tree, most of them in close association with it. Widdicombe uses a simple rule to distinguish the two by comparing the responses of an intermediate ending to inflation and deflation on the one hand with displacement of the mediastinum on the other; if the ending is stimulated more by mediastinal movement, it is classified provisionally as a mediastinal receptor. Confirmatory aids to identification would be stroking the interior of the trachea and bronchi with a catheter and irritant gases, both of which stimulate the tracheobronchial endings [Widdicombe 1954 (b)].

Natural stimulus. Since the threshold for inflation of most of these endings is relatively high, it is expected that they will not be stimulated during eupnoea but many of them will be stimulated during hyperpnoea and coughing. Therefore, unlike the slowly adapting endings their natural stimulus is not inflation of the lungs, but like the rapidly adapting ones, it appears to be local irritation of the endings, because they are stimulated by stroking the interior of the trachea or bronchi, a little by powdered talc and some, clearly by irritant gases. Ignoring adaptation indices, it would appear that the intermediate endings are more akin to the rapidly adapting ones than to the slowly adapting bronchial endings.

E. Pulmonary deflation receptors

It is now known that there are endings in the lungs which are normally inactive, but which may be stimulated occasionally by marked collapse of the lungs (Fig. 14). They are stimulated markedly and consistently by certain chemical substances notably phenyl diguanide (Fig. 13), and this excitation

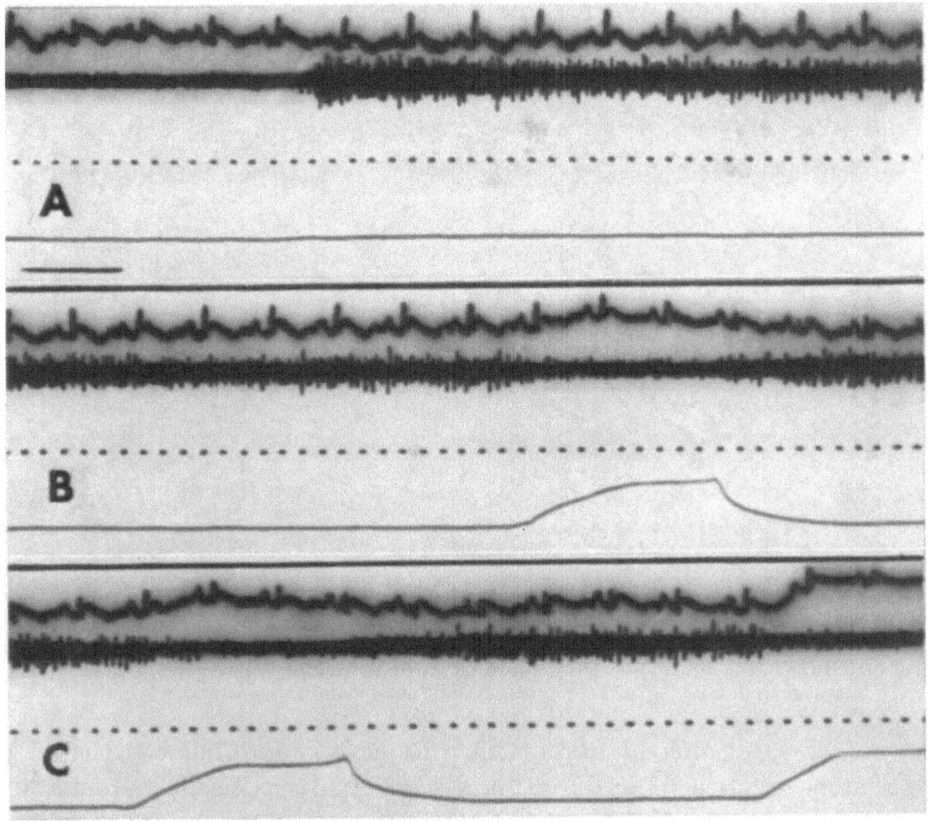

Fig. 13 A—C. Identification of deflation fibres in a cat with intact chest after atropine 1 mg/kg. A, phenyl diguanide, 175 µg, injected at signal. B and C, effects of inflation. From above downwards in each record: e.c.g.; impulses in deflation fibres; time in $^1/_{10}$ sec; intrapleural pressure, rise in pressure upwards, and injection signal in A. The three records are continuous [PAINTAL 1957 (a)]

is enhanced by deflation of the lungs and is completely inhibited by inflation (Fig. 15 B). These endings are what are now known as pulmonary deflation receptors [PAINTAL 1955 (a), 1957 (a)].

However, the existence of pulmonary receptors excited by deflation of the lungs was reported by KELLER and LOESER (1926) who recorded impulses from the whole vagus and later confirmed by ADRIAN (1933) from records of unitary discharges. ADRIAN found that the endings that could be stimulated by suction of air from the lungs but not by external compression of the chest, adapt rapidly, but that the majority were like those stimulated during inspiration. Although he suspected strongly that some of the endings that were stimulated by suction of air from the trachea were also stimulated by inflation, he nevertheless was certain that some endings were stimulated by deflation. Apparently the impulses of these fibres were similar to those arising from stretch receptors. In the rabbit a discharge during expiration could be produced by external pressure on the chest as carried out by Hammouda and Wilson (1932).

Thereafter, although Whitteridge and Bülbring (1944), Whitteridge (1948), Knowlton and Larrabee (1946), Dickinson (1950), recorded impulses

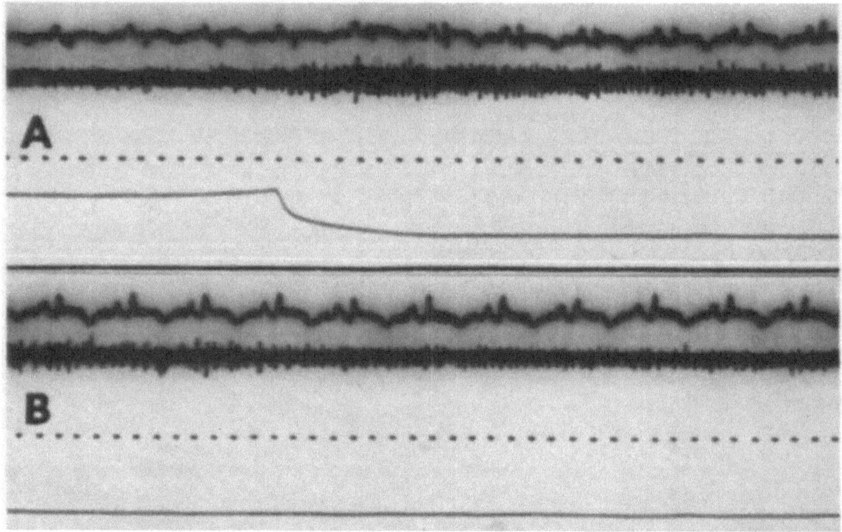

Fig. 14A and B. Activity in deflation fibres in a cat with intact chest and without drug sensitization. The cat had been atropinized. From above downwards: e.c.g.; impulses in deflation fibres; time in $^1/_{10}$ sec; intrapleural pressure, inflation upwards. A and B are continuous [Paintal 1957 (a)]

from various vagal afferent fibres with a respiratory rhythm, none of these investigators recorded impulses in fibres stimulated specifically by deflation except perhaps for some unpublished observations by Pearce and Whitteridge (cf. Whitteridge 1950). Later with some difficulty 4 fibres that could be stimulated by suction of air from the trachea but not by inflation of the lungs were isolated [Paintal 1953 (c)], but Widdicombe [1954 (b)] did not come across any. Infrequently, in the rabbit Bein and Heilmich (1949) recorded impulses during expiration.

In retrospect it may be concluded that Adrian's deflation endings were in fact the slowly adapting bronchial receptors described by Widdicombe [1954 (b)]. This may also apply to the unpublished observations of Pearce and Whitteridge (Whitteridge 1950) and to 3 of the 4 fibres described earlier [Paintal 1953 (c)]; the 4th probably arose from a deflation ending because its conduction velocity was 2.8 m/sec [Paintal 1953 (c)], a velocity that seems to be characteristic of fibres from deflation receptors (Paintal, unpublished observations).

Isolation of deflation afferent fibres. Like the endings from the stomach, the deflation receptors were found accidentally. Whereas the gastric ones were found while looking for the thoracic endings responsible for the amidine reflexes (DAWES and MOTT 1950; DAWES, MOTT and WIDDICOMBE 1951), the deflation receptors were found while studying the endings from the stomach [PAINTAL 1954 (c)]. At first all that could be said was that there were endings somewhere in the thorax because they were stimulated at a relatively short interval (Fig. 15A) following injection of phenyl diguanide into the right atrium [PAINTAL 1954 (c)]. Therefore for some time the only method available for isolating these thoracic afferent fibres was to see if injection of phenyl diguanide into the right atrium aroused impulses in silent filaments of the vagus. The abdominal endings were excluded by noting that intraaortic injections of phenyl diguanide did not arouse any impulses in the filaments. Even-

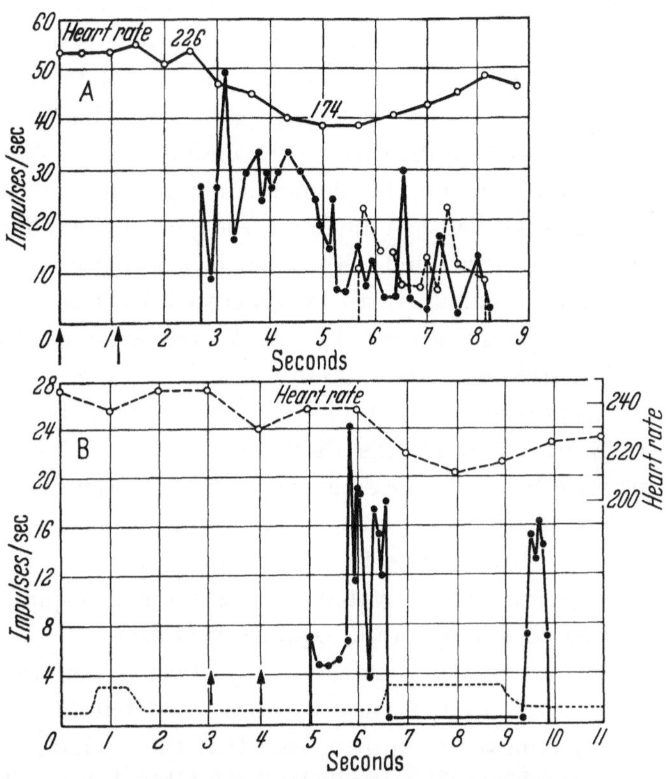

Fig. 15A and B. Activity in two deflation fibres following injection of phenyl diguanide at arrows into the right atrium. A (intact chest) shows the difference in the injection-discharge time between a deflation fibre and a gastric stretch fibre —o—o— recorded simultaneously. B shows the effect of injecting phenyl diguanide between arrows during collapse of the lungs in a cat with open chest. Note the effect on the discharge of inflation and collapse of the lungs indicated by the interrupted line, inflation upwards [PAINTAL 1955 (a)]

tually it was found that the endings were located in the lungs and that they could be stimulated by deflation of the lungs provided they were sensitized by phenyl diguanide [PAINTAL 1955 (a)].

In view of this it is now easier to isolate afferent fibres from deflation receptors and the procedure followed is to dissect a filament containing no active fibres and in which no impulses are aroused by injecting phenyl diguanide into the abdominal aorta, but in which impulses are aroused with a short latency, i.e. less than 3 sec, by injecting the substance into the right atrium (Figs. 13, 15); this discharge should be reduced or abolished by inflating the lungs and increased after deflation (Figs. 13, 15 B) [PAINTAL 1957 (a)]. In rare instances it is possible to obtain a discharge by manual compression of the chest after an inflation without any prior sensitization by phenyl diguanide, and more rarely, a discharge may be produced by passive collapse of the lungs following inflation with intact chest (Fig. 14). Using these procedures it is thus possible to isolate deflation fibres with some certainty without opening the chest, but there are fibres in which there is no effect of pressing the intact chest; in such cases it is possible to stimulate the endings by allowing the lungs to collapse completely after opening the chest; such a procedure most often succeeds in identifying the ending beyond doubt.

Natural stimulus for deflation receptors. The best stimulus for deflation receptors is collapse of the lungs, the greater the collapse, the larger the number of impulses generated. However, in the majority of endings, not more than about 1—3 impulses can be generated by this stimulus in the absence of chemical sensitization. The peak frequency of discharge does not exceed 50 impulses/sec and is usually less than 30/sec (Fig. 15 A and B). Although in exceptional cases a train of impulses lasting about 1—3 sec can be produced following deflation, the discharge is nevertheless rapidly adapting. In no instance has a slowly adapting discharge been produced. Repeated inflation and deflation most often weakens the response. Since complete collapse is necessary to stimulate the endings, it is almost certain that in the cat, these endings are not activated during normal respiration, but they could be stimulated during forced expiratory efforts.

In view of the weak excitatory effect of deflation, the role of deflation as the natural stimulus may be questioned but the doubts concerning this are largely removed when it is seen that the discharge elicited by collapse during drug sensitization is completely inhibited by inflation (Fig. 15 B). These effects are primarily mechanical and are not due to vascular changes because the endings can be stimulated by collapse of the lungs after cardiac standstill [Paintal 1955 (a)].

The endings are not stimulated by oxygen lack or CO_2 excess or both since they are not stimulated by asphyxia or by haemorrhage. On the other hand they are stimulated by congestion of the lungs produced by occluding the left atrio-ventricular junction partially. The degree of stimulation is much less than that produced by deflation and the discharge is of low frequency [Paintal 1955 (a)]. This has been confirmed in recent experiments on single deflation endings (unpublished observations). It has also been found that this excitation by congestion is diminished by repeated trials; there is a suspicion that this may be due to loss of sensitivity of the ending.

The impulses in these fibres are the smallest that have been recorded in vagal afferent fibres, some of them barely exceeding the noise level of the pre-amplifier. The largest ones have an amplitude of 25 μv. So far there are no published records of discharges in a single deflation fibre owing to the difficulty of subdividing thin filaments without injuring the fibres. This contrasts with gastric stretch fibres which can be isolated as single units with a little practice even though these fibres are non-medullated (IGGO 1958). Perhaps this is due to the possibility that the deflation fibres are the very thinnest myelinated fibres with conduction velocities between 2 and 6 m/sec (unpublished observations). It has been noted repeatedly that the deflation fibres tend to be aggregated together in the cervical vagus. Thus it is sometimes possible to isolate a strand consisting exclusively of about 20 deflation fibres, these being distributed to each one of the filaments in the subdivisions of the strand [PAINTAL 1955 (a)].

Location of the endings. The main evidence concerning the location of the deflation receptors has come from the responses of the endings to various chemical substances. In the initial investigation [PAINTAL 1955 (a)] when the chief interest centred on the injection-discharge times following phenyl diguanide, efforts were directed at explaining where the endings must be located in order that they may be stimulated within 0.9—2.7 sec after injections of phenyl diguanide. It was shown that the endings could not be located in the pulmonary artery, its bifurcation or its main branches as far as the hilum of the lung. After making certain allowances in the results of GRAY and PATON (1949) it was concluded that the endings were probably located in the pulmonary capillaries since it was estimated that the pulmonary circulation time would be about 2.5 sec [PAINTAL 1955 (a)].

Subsequent work showed that this figure for pulmonary circulation time was a gross overestimate because it was found that the start to start time (which is what one is concerned with in a correlation with injection-discharge times) using P_{32} ranged from 0.6—1.5 sec (average = 1.0 sec) [PAINTAL 1956 (a)]. This is the time between the injection of a slug of P_{32} into the right atrium and its arrival at the left atrium. This has been confirmed in more extensive experiments with better equipment by GOEL (personal communication).

In a later investigation on deflation receptors it was demonstrated that some time elapses between the arrival of the chemical substance at the ending and the beginning of excitation because of two reasons. First, larger doses of phenyl diguanide stimulate the endings with a shorter latency and second, different substances yield different injection-discharge times, those following nicotine and urethane being the lowest (average = 1.0 sec in both) [PAINTAL 1957 (a)]. Since the pulmonary circulation time is about 0.8 sec, even these low figures for injection-discharge times are large enough to allow one to consider that the endings may be located anywhere between the pulmonary arterioles and the pulmonary veins.

Equally valuable information about the location of the endings is provided by the responses of the endings to volatile anaesthetics [PAINTAL 1957 (a)].

After making allowances for the dead space in the lungs, and probable total conduction time from the endings to the recording electrodes, it was found that the interval between the likely arrival of ether vapour in the alveoli and the excitation of the ending was 0.16 sec. This figure is probably high because recent experiments (unpublished) reveal that the allowance made for total conduction time was smaller than it actually is. These experiments have therefore confirmed the suggestion put forward earlier, namely that the endings must be located in a region where they are directly accessible by ether vapour and also accessible by chemical substances through the pulmonary circulation. The only part of the lungs where this is possible is the respiratory bronchioles and the region distal to them [PAINTAL 1957 (a)].

X. Cardiovascular afferent fibres

A. Systemic arterial baro- and chemoreceptors

These endings have been dealt with so exhaustively by HEYMANS and NEIL in their recent book (1958) that little else can be added. Although in this book much of the information concerning the properties of the endings relate to baro- and chemoreceptors of the carotid artery, it can be applied directly to the endings in the aorta and the region of the brachiocephalic trunk where BIANCONI and GREEN have recently found additional baro- [BIANCONI and GREEN 1959 (a)] and chemo- (BIANCONI and GREEN 1960) sensory areas.

The conduction velocities of the arterial baroreceptors are given in Table 3 and their distribution shown in Fig. 3. Since the conduction velocities of only a few afferent fibres from chemoreceptors are known (Table 1) a frequency distribution has not been made. However, the conduction velocities of several more fibres have been determined recently (unpublished observations) and these show that the lower limit of conduction velocities is less than that reported (Table 2).

B. Atrial receptors

It is now well known that there are two main types of atrial endings, the type A and type B [PAINTAL 1953 (a); WHITTERIDGE 1953; STRUPPLER 1955; HENRY and PEARCE 1956; HEYMANS and NEIL 1958; PAINTAL 1962]. Each type has its own characteristic pattern of discharge under ordinary experimental conditions with intact chests and the animal breathing spontaneously. The characteristic feature of the type B endings, earlier known as pulmonary vascular receptors (WHITTERIDGE 1947, 1948, 1953), is that they have a late systolic burst of impulses coincident with the v wave of the atrial pressure curve (Figs. 16, 18, 19); characteristically they do not have any impulses during atrial contraction. This silence during atrial contraction distinguishes them from the type A endings whose most characteristic feature is that they have a relatively high frequency burst of impulses coincident with the *a* wave of the atrial pressure curve (Figs. 22, 23). Most type A endings have only this burst, but there are some endings currently designated as type A in which there is a prominent burst of impulses in another part of the cardiac cycle as well, notably in time with the *v* or *c* waves of the atrial pressure curve [WHITTERIDGE 1948, 1953; DICKINSON 1950; PAINTAL 1953 (c); HEYMANS and NEIL 1958]. Rarely one may come across atrial receptors with only an early systolic burst of impulses [PAINTAL 1955 (b); LANGREHR 1960 (a)].

1. Type B atrial receptors

The criteria for identifying type B endings with closed chest have been described [WHITTERIDGE 1948; PAINTAL 1953 (a)]. Some of the criteria have been summarised in Table 3. However, the identification of the endings depends primarily on their having a late systolic burst of impulses which distinguishes them from the arterial baroreceptors that are characterised by the presence of a midsystolic discharge. In cases of doubt other criteria can be of considerable help. Thus the discharge in type B endings as a rule shows marked respiratory fluctuations and the frequency of discharge is rarely more than 100 impulses/sec [PAINTAL 1953 (a)].

While it may be relatively easy to identify type B endings from others, identification of type B endings in the right atrium from those in the left may sometimes present difficulties. The ways in which identification of the two with closed chest may be attempted are summarised in Table 3. Thus the discharge increases early in inspiration in right-sided endings and is later in onset in left-sided ones. This agrees with the differences in right and left atrial filling (CAHOON, MICHAEL and JOHNSON 1941). Again, in right atrial endings the discharge sets in immediately after releasing the inflation of the lungs; this happens one to two cycles later in left atrial endings. In any case it is not possible to be certain about the location of any particular ending unless the chest is opened and the ending is localized to a particular chamber of the heart [PAINTAL 1953 (a); COLERIDGE et al. 1957; LANGREHR 1960 (a)]. As a rule this is easily accomplished by simply noting the effect of occluding the pulmonary artery. If the ending is located in the right atrium, the discharge will increase markedly, and if in the left, it will fall and disappear. Further confirmation can then be obtained by occluding the *a-v* junctions and by mechanical stimulation which provides unequivocal information if the stimulation is carried out after cardiac standstill.

The pattern of discharge in most type B endings is as shown in Figs. 16, 18 and 19, i.e. the frequency of discharge gradually increases to reach a peak and then falls, a sort of crescendo pattern quite unlike what is frequently seen in systemic and pulmonary arterial receptors in which the peak frequency is attained initially and this starts to decline from the very beginning (Fig. 26).

The same is true in the dog (Henry and Pearce 1956; Mühl, Scholderer and Kramer 1956) and in monkeys (Chapman and Pearce 1959). However, the picture is variable, but in the same ending although the number of impulses in each cycle may vary considerably during different phases of respiration, the pattern of discharge as shown by Whitteridge (1948) often remains essentially the same (Fig. 16).

The pattern of discharge and the amount of impulse activity may change markedly after opening the chest, but it is remarkable that in spite of this radical procedure and the pericardium being slit, the pattern of discharge in many fibres retains its original features (Figs. 18, 19). In some endings other bursts of impulses may appear, notably the *a* burst especially if much blood has been lost as shown in Fig. 21 (Pearce, Henry and Chapman 1956). The position of the heart and the presence of ligatures also makes a difference and the pattern of discharge can be altered considerably. Thus as shown in Fig. 21 after haemorrhage, there may be only an *a* burst of impulses! It is therefore important that identification of a particular ending be done under control conditions with the

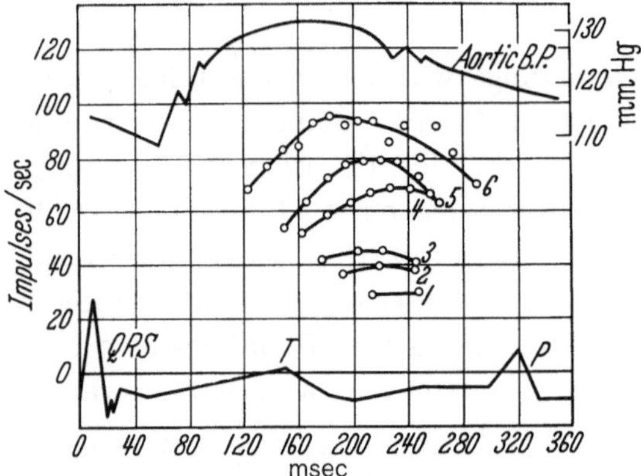

Fig. 16. Superimposed records of the activity in a fibre from a type B atrial receptor during successive stages of inspiration, 1—5, and during suction of air from the trachea, 6. A single representative aortic blood pressure curve and an e.c.g. are also shown, though minor changes in duration of the cardiac cycle and in the amplitude of the pressure pulse occurred with respiration (Whitteridge 1948)

chest intact and the original identification be retained although the pattern may alter so much as to make it characteristic of another type of ending after some manoeuvre (e.g. opening the chest, haemorrhage, etc.). If this is not done confusion will arise.

The pattern of discharge may also change radically if the sequence of events in the heart changes. Thus if the heart acquires a nodal rhythm there may be a prominent burst, the so-called early discharge early in systole [Paintal 1955 (b)]. This is due to the atrium contracting against closed *a-v* valves so that apart from back flow into the veins, the volume of the atrium does not fall and the contraction therefore stimulates the ending.

Natural stimulus for type B endings. In the initial investigation it was shown that the endings were slowly adapting stretch receptors stimulated by increased atrial filling [Paintal 1953 (a)]. This was clearly demonstrated in experiments on left atrial endings in *isolated in situ* left atria. The left atrium was isolated from the rest of the heart by clamping the pulmonary veins and the left *a-v* junction so that it became a closed chamber. Injecting graded amounts of fluid yielded graded activity in the ending which was linearly related to the amount of filling. Such experiments also showed that the endings were slowly adapting, the frequency of discharge being maintained for some time after the initial fall in frequency following the end of stimulus increase. This has been confirmed in similar experiments in dogs by MÜHL (see KRAMER 1959) and with graded distensions of the left atrium by balloons (HENRY and PEARCE 1956; PEARCE, HENRY and CHAPMAN 1956). It was therefore concluded that the normal stimulus for the type B endings was increased atrial filling [PAINTAL 1953 (a)]. According to this view the discharge starts when the filling threshold of the ending is attained (on the rising slope of the v wave) and it falls when the atrium empties after the opening of the a-v valves on the downslope of the v wave. This is quite clear in the case of those endings where the burst ends before the P wave of the ECG (Figs. 18, 19 and 20).

In this connection, the v wave of the atrial pressure curve has been used as a rough guide to the amount of blood in the atrium and it was found that the discharge ran parallel to the shape of the v wave [PAINTAL 1953 (a)] but owing to defective pressure records which showed that there was marked deviation of the discharge from the shape of the v wave in some cases, this aspect was not pursued further. However, later in experiments on dogs HENRY and PEARCE (1956) brought out this relationship and there is now no doubt that satisfactory pressure records show that the discharge in the fibres parallels the v wave, in some very closely (Figs. 18, 19, 20).

It is conceivable that the results obtained under unusual conditions as with isolated *in situ* atria may not be applicable under normal conditions and it is therefore necessary to get records of impulse activity simultaneously with records of phasic changes in atrial volume. This is not possible at present because there is no reliable method of recording phasic changes in atrial volume. The atrial diastolic pressure has therefore been used as a guide to the amount of atrial filling since the compliance of the atrium may be assumed to be constant during atrial diastole and atrial pressure varies almost linearly with atrial filling at levels of filling normally encountered (OPDYKE, DUOMARCO, DILLON, SCHREIBER, LITTLE and SEELY 1948; LITTLE 1949, 1960; IRISAWA, GREER and RUSHMER 1959). As expected, the activity in the endings is found to be closely related to atrial volume (Fig. 17). This is especially so if the various factors influencing the activity in an ending are taken into consideration. Thus the beginning of the discharge will depend on the threshold of atrial filling [PAINTAL 1953 (a)] and the frequency will depend on the rate of increase of filling. The total number of impulses in a burst will therefore depend on the threshold of filling, the rate of filling, and the total duration of increased filling which is determined by the moment the a-v valves open. In the final calculation it is found that the discharge is related to the difference between peak filling and threshold filling because the time factor cancels out when the rate of filling is multiplied by the duration of increased filling (Fig. 17) (Paintal, unpublished observations).

In contrast Langrehr [1960 (b)] believes that the activity in type B endings is determined by ventricular activity. This conclusion is based largely on a correlation between the activity in the endings and stroke volume. Recent experiments have failed to confirm this hypothesis. Thus when activity in the fibre is recorded simultaneously with atrial and ventricular pressure, it is found that the activity follows the diastolic atrial pressure and is often totally unrelated to ventricular pressures (Fig. 18). However, the best evidence that the activity in type B endings is not directly influenced by the strength of ventricular contractions is obtained from the effects of premature ventricular contractions (Fig. 19). These results show consistently that the activity in the ending is much greater during premature ventricular contractions which are weaker than normal contractions (Wiggers 1928). Actually the weaker the contraction the greater is the discharge in the ending. But this is a spurious relation because of the direct relation between the amount of atrial filling and the discharge in the ending. If a premature ventricular contraction occurs soon after a normal one when the atrium has had no chance to empty itself, most of the blood will be retained in the atrium. This will stimulate the ending markedly and will also result in a very weak ventricular contraction. This is shown in Fig. 19 which demonstrates clearly that the discharge is closely related to the atrial pressure curve and that it cannot be related to ventricular activity. However, in this figure there is no record of ventricular pressure but there is one in Fig. 18 which proves that the activity in type B endings is not related directly to the strength of ventricular contraction at all. However, there can be an indirect relation because ventricular contraction can influence the amount of blood entering the atrium (Brecher 1958). It is therefore not suprising that Langrehr [1960 (b)] found a close correlation between stroke volume and impulse activity. Besides, stroke volume will also depend on atrial filling. For similar reasons the

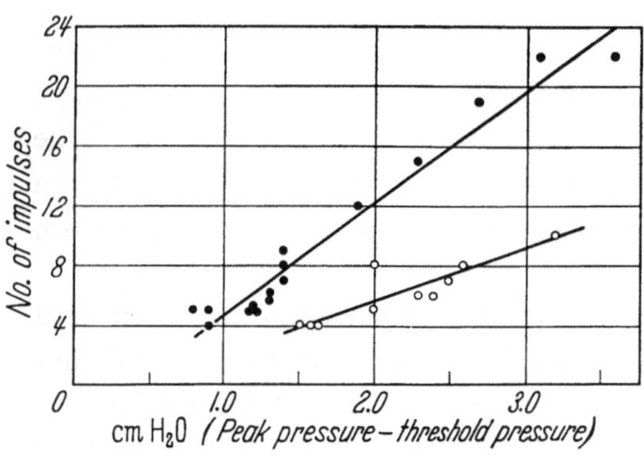

Fig. 17. Graphs showing that the discharge in two left atrial type B receptors is related to atrial volume. Although the abscissa represents the difference between peak pressure of the v wave and threshold pressure, it also represents the difference between peak volume and threshold volume of the atrium since pressure is a function of volume. Actually abscissa is equal to average rate of filling × duration of filling (see text). Chest was open in both these experiments (Paintal 1962, unpublished observations)

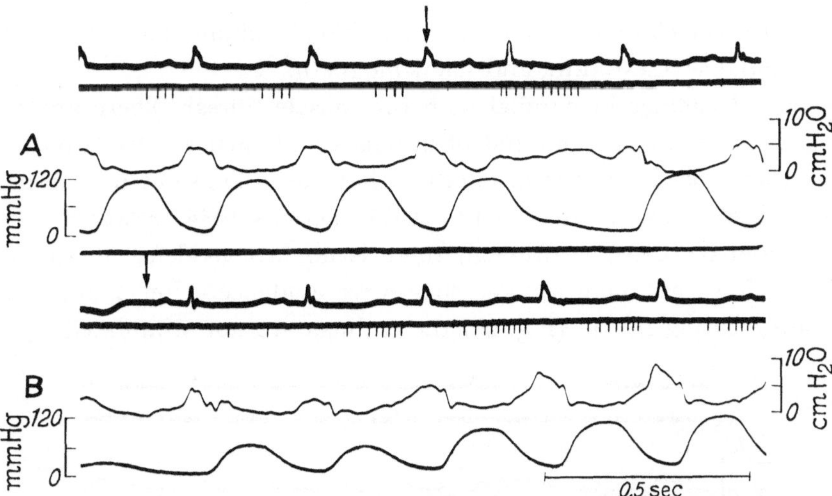

Fig. 18 A and B. Records of activity in a left atrial type B fibre after opening the chest showing that the discharge is not related to ventricular activity. The pulmonary artery was occluded at arrow in A and released at arrow in B. Increased activity in the fibre appeared in the third cycle in B when left ventricular pressure is low; in fact the pressure is lower than in the second cycle. Note also that the fibre is silent in the last cycle in A when ventricular pressure is highest. The increased activity runs parallel with increased atrial diastolic pressure. From above downwards in each record, e.c.g.; impulses in the fibre; left atrial pressure and left ventricular pressure. Total conduction time in the fibre was 11 msecs (PAINTAL 1962, unpublished observations)

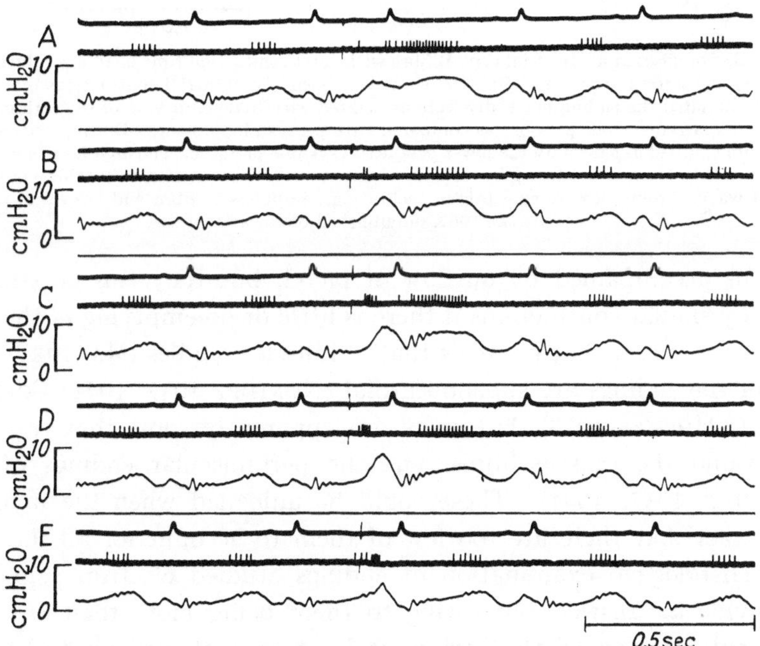

Fig. 19 A—E. Records of impulses in a left atrial type B fibre showing the effect of premature ventricular contractions produced by stimulating the atrium electrically in different parts of the cardiac cycle. In every case, premature contraction is associated with considerable increase in activity which corresponds closely to the rise in left atrial pressure at this time. This increase is greater, the earlier the premature contraction, and so it falls from A to E. Activity in the fibre is therefore not related to ventricular activity at all. Note that normal activity in the fibre ends before atrial contraction begins, in fact before the P wave. From above downwards in each record, ECG; impulses in the fibre, and left atrial pressure. Total conduction time in the fibre was 10 msec. (PAINTAL 1962, unpublished observations)

existence of a correlation between thoracic blood volume and atrial activity is understandable (Langrehr and Kramer 1960).

Are type B endings in parallel with the muscle fibres? There are several type B endings in which the burst of impulses ends just as atrial contraction begins so that it has led to the speculation that the endings might be in parallel with the muscle fibres (Whitteridge 1953; Dawes 1954; Struppler 1955; Heymans and Neil 1958) although there is no firm evidence in support of this view. However, there is some suggestive evidence. For example, in arrythmically beating hearts (Fig. 20) the endings are not stimulated by atrial

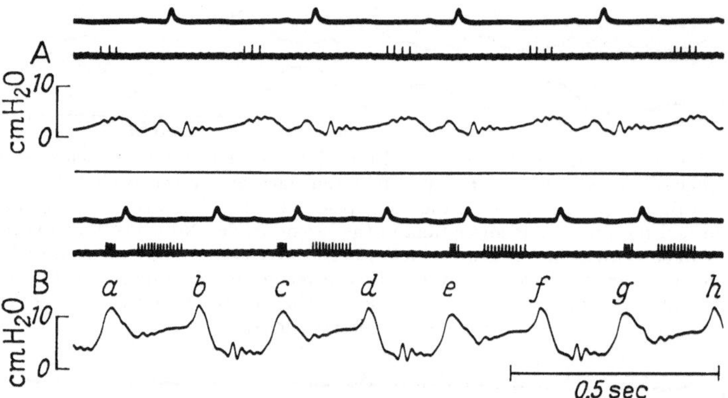

Fig. 20 A and B. Impulses in a left atrial type B fibre with total conduction time of 10 msec during normal heart beats (A) and during arrythmia (B). The chest was open. In B, atrial contractions marked *a, c, e* and *g* which stimulated the ending markedly were associated with little emptying of the atrium as shown by the raised post-atrial systolic pressure as compared with presystolic pressure. During contractions *b, d* and *f* the atrium emptied as shown by the lower post-atrial systolic pressure. The high frequency discharge which is comparable to that produced by ectopic contractions in Fig. 19 is attributable largely to the enhanced *a* wave. From above downwards in each, e.c.g.; impulses in fibre and left atrial pressure (Paintal 1962, unpublished observations)

contractions accompanied by outflow of blood, but they can be stimulated markedly by similar contractions if there is little or no emptying of the atrium (Fig. 20), a behaviour that parallels that of muscle spindles (Matthews 1933).

Most of the endings are subendocardial (Nonidez 1937, 1941; Coleridge et al. 1957; Holmes 1957), but there are apparently some that seem to be twined around the muscle fibres, viz. the perimuscular endings of Nonidez (Nonidez 1937, 1941). These could be unloaded when the myocardial fibres contract, but there are too few of them to account for all the type B endings. Histological examination of endings studied electrophysiologically might provide an answer, but owing to there being more than one ending in a particular region of the atrium it is at present difficult to be certain that the ending from which impulses were recorded was the same as that examined histologically. However, investigations along lines done by Coleridge et al. (1957) are needed so that the arrangement of the type A and type B endings is determined and a structural basis provided for the differences in the normal behaviour of the two types of endings.

Appearance of *a* impulses in type B endings. Under certain conditions [see PAINTAL 1953 (a)] such as after haemorrhage (HENRY and PEARCE 1956; PEARCE et al. 1956) a few impulses usually 2 to 3 may appear in time with the *a* wave giving the impression of an *a* burst (Fig. 21). However, most often the frequency of these impulses is much lower than that occurring in the prominent *a* burst of type A receptors. Although these impulses are usually inconstant and transitory, they nevertheless attract one's attention because they give the impression that the pattern of discharge has changed into that typical of type A endings leading some to argue [e.g. LANGREHR 1960 (a)] that there is no difference between type A and type B endings. Among one of the causes for the appearance of an *a* burst is the delayed peak of the *v* wave which in some cases may run into the *a* wave of the venous pressure curve [WHITTERIDGE 1948; PAINTAL 1953 (a)]. In such situations a knowledge of total conduction time from the ending to the recording electrodes is indispensable for determining whether the burst is due to the *v* wave or to atrial contraction.

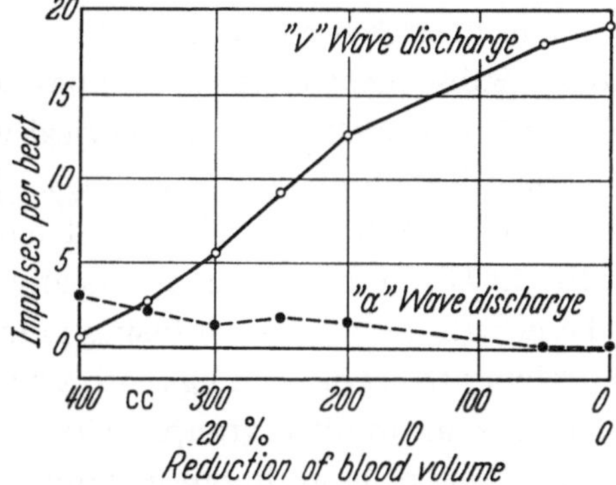

Fig. 21. Effect of haemmorhage on a type B atrial receptor of a dog. Note that the receptor had only a *v* burst of impulses, with 19 impulses per burst under normal conditions. Reduction of blood volume by 200 ml reduced the number of *v* impulses to 13 and there appeared 1 to 2 *a* impulses per cycle. Eventually reduction of 400 ml of blood led to almost complete disappearance of *v* impulses. Instead there were now 3 *a* impulses giving the impression that the pattern of discharge had changed into that of a type A receptor (CHAPMAN 1956, unpublished observations)

Location of the endings. So far no type B endings have been found in the atrial appendage. Out of 10 endings isolated initially, some were located by means of punctate stimulation in the posterior wall of the atria and some near the opening of the pulmonary veins [PAINTAL 1953 (a)]. Similar results were obtained by COLERIDGE et al. (1957) and LANGREHR [1960 (a)] in the dog.

2. Type A atrial receptors

The type A atrial endings are those which have a prominent burst of impulses in time with the *a* wave of the venous pressure curve in addition to any other burst of impulses [PAINTAL 1953 (a)].

AMANN and SCHAEFER (1943) were the first to record impulses from these receptors in the thin cardiac branches of the vagus nerve. They were originally known as venous endings (WALSH and WHITTERIDGE 1945; WHITTERIDGE 1947, 1948; DICKINSON 1950) and were later labelled as type A in order to distinguish them from the type B also located in the atria [PAINTAL 1953 (a)]. WALSH and WHITTERIDGE (1945), WHITTERIDGE (1947, 1948) and SCHAEFER (1950) related the activity in these endings to the pressure waves in the venous pulse. Apparently Jarisch and Zotterman (1948) and Neil and Zotterman (1950) also noted that atrial pressure played a part in stimulating the endings, but because they seemed to have been more impressed by the fact that the main burst of activity in these endings survives during occlusion of both caval veins, they felt that the endings must be excited by the movements of the heart.

Like Amann and Schaefer (1943), Jarisch and Zotterman recorded impulses from undissected thin branches of the vagus going to the auricles, which as their records show contain many active fibres, including some from pulmonary stretch receptors so that it is difficult to draw unequivocal conclusions. Subsequently, these endings were studied chiefly by recording from filaments dissected from the cervical vagus (see Heymans and Neil 1958).

Identification of type A endings. It is customary to identify a fibre as arising from a type A ending if it has a burst of impulses after the P wave of the ECG whether or not it has any other burst of impulses. This criterion is beset with pitfalls, is confusing practically, and all that it succeeds in doing is dividing the atrial endings into two categories, the type B in one group and all others in another group. The confusion arises when a type B receptor acquires one or more impulses after the P wave of the ECG, which happens, e.g. after haemorrhage (Fig. 21) (Pearce et al. 1956), and the fibre is isolated when this has already happened or when a type A receptor acquires one or more v impulses. In order to avoid any confusion it is suggested that no attention should be paid to the appearance of only one impulse or two impulses if these are present in only a few cycles. Secondly, only those endings should be labeled as type A which have *only* an *a* burst of impulses, barring occasional impulses in other parts of the cycle. This necessitates the creation of a third category, the intermediate type which are fortunately much fewer and which have both *prominent a* and *v* bursts of impulses and less frequently a *c* burst as well. It is important that the identification be carried out under control experimental conditions with the chest intact and the animal breathing spontaneously. The intermediate type are almost certainly not a separate functional group but merely extreme variations of type B or type A endings and it is usually possible to say whether such an ending behaves like a type B or a type A receptor.

For differentiating right atrial type A endings from left atrial ones the procedure described in connection with type B endings can be used. But differentiation is clearly more difficult than with type B endings. This is because type A endings show little fluctuation in activity. Usually it is necessary to open the chest and locate the ending as described already. It is important to remember that there are some pulmonary stretch receptors with a cardiac rhythm and a prominent *a* burst of impulses [Coleridge et al. 1957; Bianconi and Green 1959 (c)] and it is therefore necessary to exclude these.

There are also some with typically type B pattern of activity [BIANCONI and GREEN 1959 (c)].

Natural stimulus for type A endings. Ever since the demonstration that the venous fibres having a, c and v bursts of impulses may follow closely the respective waves in the venous pressure curve (WHITTERIDGE 1947, 1948) it has been largely accepted that the type A endings are stimulated by rise in atrial pressure which is their natural stimulus; increase in pressure increases

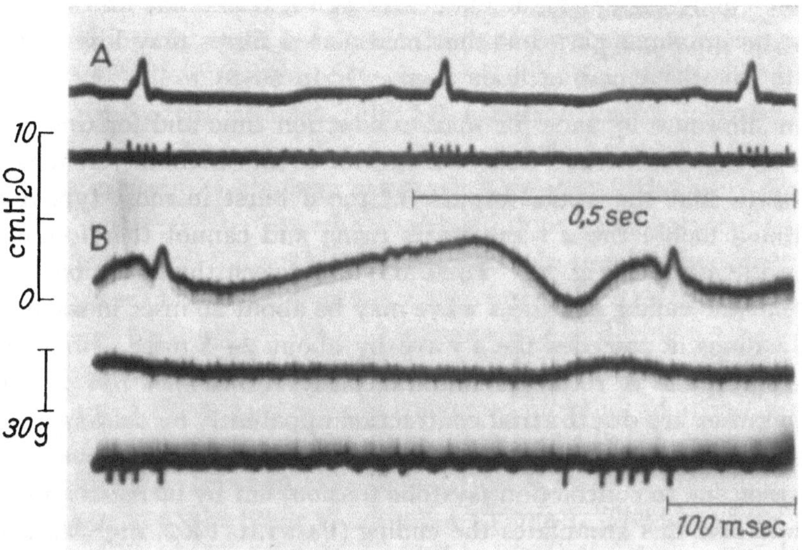

Fig. 22A and B. Impulses in a left atrial type A fibre with total conduction time of 12 msec. A, is a section of the continuous record of e.c.g. and impulses in the fibre. B, is a sweep of from above downwards, left atrial pressure; left atrial contraction and impulses in the fibre. The first burst of impulses in the sweep corresponds to the first burst in A. The first impulses of each burst is attributable to atrial contraction and not the a wave because it appears before the a wave (PAINTAL 1962, unpublished observations)

their activity. Further, the fact that the main burst of impulses after the P wave of the ECG is always the most prominent fits in with this view because peak pressure is attained during the a wave (WIGGERS 1928). The correlation between atrial pressure and the activity in the endings is good (DICKINSON 1950) but as noted by DICKINSON, there may be certain exceptions, some of which may be explained if it is assumed that the endings adapt rapidly. Actually, rapid adaptation is implicit because in the cat most commonly one comes across, type A endings which do not have any v impulses, often in spite of a prominent v wave. If the endings are exclusively stimulated by rise in atrial pressure this must mean that they adapt rapidly. STRUPPLER and STRUPPLER (1955) extended these observations and concluded that the type A endings were stimulated by rate of rise of pressure and not merely by the magnitude of the pressure. This conclusion follows from the knowledge that the endings adapt rapidly and it arises from the need to explain why there may be very little or no activity in type A endings although the peak pressure of the *v* wave may be only a little lower than that in the *a* wave.

In all these studies it has been assumed that the *a* burst of impulses owes its origin to the *a* wave especially since the two are very often coincident, it being usually assumed that the total conduction time from the ending to the recording electrodes is not of practical significance. This is not so (PAINTAL 1962). That such an assumption is unsafe is shown by the observation that the conduction velocity in the distal part of vagal afferent fibres may be less than in the proximal part and that medullated fibres may lose their myelin sheath in the distal part of their course (IGGO 1958).

If an allowance is made for total conduction time and for time for transmission of the pulse wave from the atrium to the manometer, it is easy to demonstrate that the initial impulse of the *a* burst in most type A endings are initiated before the *a* wave starts rising and cannot therefore owe their origin to the *a* wave (Fig. 22). The interval between the initiation of the first impulse at the ending and the *a* wave may be about 20 msec in some cases; in several endings it precedes the *a* wave by about 2—5 msec. However, when atrial contraction is recorded simultaneously it becomes obvious that the initial impulses are due to atrial contraction apparently by causing an increase in tension in the atrial wall. It appears that both increase in pressure (*a* wave) and tension due to contraction (systolic tension) act by increasing the tension in the wall and this stimulates the ending (PAINTAL 1962, unpublished observations).

Neither systolic tension nor increase in pressure are indispensable stimuli for the type A endings. Thus when the *v* wave is made to increase considerably, e.g. by occluding the pulmonary artery in the case of right atrial receptors, the ending is stimulated during the *v* wave purely by increase in atrial pressure. On the other hand, the pressure may be drastically reduced and the endings still continue to fire [JARISCH and ZOTTERMAN 1948; LANGREHR 1960 (a)]. However, under such conditions although the atrial diastolic pressure is greatly reduced, the amplitude of the *a* wave may not be reduced at all. In any case proof that the endings can be stimulated without any pressure in the atrium is provided by the observation that the endings yield bursts of impulses with each atrial contraction although the *a-v* junction may be widely slit and the atrium cut open and the blood allowed to flow out freely. Here, the only stimulus present is systolic tension. As a matter of fact if a strip of the atrium containing a receptor is held under some tension, it will yield impulses with each contraction (PAINTAL 1962, unpublished observations).

Although it is true that pressure is not an indispensable requirement for stimulating the endings, this fact should nevertheless not be allowed to overshadow its role in stimulating the type A endings normally because there is much evidence to show that it plays an important part normally and may even be the dominant stimulus. Thus in most endings when the *a* burst is compared with the development of atrial contraction and the *a* wave, it becomes apparent that most of the later impulses of the *a* burst must owe their origin to the *a* wave.

Appearance of *v* impulses in type A receptors. Since the appearance of *v* impulses depends on the *v* wave (WHITTERIDGE 1948) the simplest way of introducing *v* impulses is to increase the peak pressure of the *v* wave by increasing atrial filling. With open chest this is easily accomplished by partially preventing the outflow of blood, e.g. by occluding the pulmonary artery in the case of right atrial endings (Fig. 23) (JARISCH and ZOTTERMAN 1948; PAINTAL, unpublished observations). With closed chest *v* impulses can be introduced by suction of air from the trachea or after inflation of the lungs. Other causes relate to the experimental conditions such as attachments of ligatures, catheters, and the position of the heart. In some endings the appearance of a prominent burst of *v* impulses is associated with the reduction or even disappearance of the *a* burst. When this happens, e.g. during occlusion of the pulmonary artery in the case of right atrial endings, it gives the impression that the pattern of discharge has been converted into that of a type B ending. Accordingly the argument, which will be discussed below, is advanced that there is no difference between type A and type B endings and that one can take on the pattern of the other [LANGREHR 1960 (a)]. The loss of the *a* burst may be due to reduction in systolic tension, reduction in the amplitude of the *a* wave or if these are unchanged to the "off effect" (KATZ 1950) generated by the exaggerated downslope of the enhanced *v* wave. It is not due to the depression caused by the orthodromic discharge of *v* impulses themselves because a train of antidromic impulses depress the ending only a little. Actually, the only effect of antidromic impulses arriving at the ending at a time that *v* impulses usually appear is to cause a delay of about a couple of milliseconds in the appearance of the first impulse of the *a* burst (PAINTAL 1962, unpublished observations).

Function of type A receptors. It is generally recognised that the type B receptors are stretch receptors and that their function is to signal the amount of atrial filling [PAINTAL 1953 (a); WHITTERIDGE 1953; HENRY and PEARCE 1956; GAUER and HENRY 1956; KRAMER 1959; HEYMANS and NEIL 1958]. In contrast the position of the type A endings is quite undecided. However, three possibilities can be considered. First, that the type A endings are, like the type B, also stretch receptors and that their function is essentially similar to that of the type B. This concept is the kind one would arrive at from LANGREHR's conclusions [LANGREHR 1960 (a)]. On this basis the absence of *v* impulses normally in type A endings could be explained by assuming that the endings have a high threshold to stretch and the fact that *v* impulses may appear or be made to appear when atrial filling is increased would support the assumption. Further, as shown in Table 1 and Fig. 6 the conduction velocities of their fibres are similar.

There are two objections to this view. Firstly, although it is likely that they are stretch receptors, there is little justification for saying that they signal changes in atrial volume because it is a common experience to find that many endings show little or no respiratory fluctuations in the discharge (Table 4), quite unlike the marked respiratory fluctuations seen in type B endings which are due to the varying amount of inflow of blood during respiration (Cahoon, Michael and Johnson 1941; Opdyke, van Noate and Brecher 1950). The striking difference between the two types of endings as far as this point is concerned is brought out most clearly when impulses are recorded simultaneously from type A and type B endings located in the same chamber; the type B will show the characteristic fluctuations with variations in venous inflow, and the type A (i.e. those with only an *a* burst) will show hardly any variation at all (Paintal, unpublished observations).

The second objection to the view that type A endings are functionally similar to the type B is that, if this is true, then there should be various gradations between the pure type B and the pure type A and there should be some sort of normal distribution with the pure types at the two ends of the distribution curve. This is not the case. There are many more pure type B and type A endings (if one ignores a solitary impulse in a part of the cycle, see p. 118). In fact out of a total of 22 fibres from the atrium isolated at random, there were 9 pure type B and 10 pure type A. One ending had a low frequency burst of 2 *a* impulses and a weak *v* burst and this could be attributed to considerable loss of blood that had occurred prior to isolating the fibres. Two other fibres with prominent *a* bursts had 2—4 *v* impulses in a few cycles after release of inflation of the lungs, none during normal respiration (Paintal, unpublished observations).

The second possibility is that the type A endings signal the level of intra-atrial pressure, i.e. they are the kind of pressure receptors found in the aorta and carotid arteries. In support of this is the evidence that there is a linear relation between pressure and impulse activity (Dickinson 1950). Because some of them adapt rapidly (Dickinson 1950), they could also be thought of as signaling the rate of change of pressure (Struppler and Struppler 1955). Against this hypothesis is the striking fact that the discharge in the endings may not be reduced at all although both the diastolic pressure and peak pressure of the *a* wave may fall considerably when blood flow into the particular chamber concerned is prevented (Jarisch and Zotterman 1948; Paintal 1962, unpublished observations). In fact the endings may continue to discharge although there is no blood in the atrium (see p. 120).

The third possibility is that the endings signal the amount of tension in the atrial wall. The increase in tension may be produced by atrial systole or by increase in pressure. Both these factors will increase the tension in the wall. The absence of the *v* burst is explained by the tension in the wall being sub-threshold for the ending normally. This will increase when pressure increases;

Table 3. *Summary of characteristics useful in identifying various types of cardiovascular afferent fibres* [PAINTAL 1955 (b)]

Type of receptors	Main volley of impulses	Effect of inspiration on impulse activity	Effect of inflation on impulse activity	Return of activity after inflation	Response to "veriloid"
Right atrial type A	Presystolic	Increased early in inspiration	Reduced or abolished	Early	Not stimulated
Right atrial type B	Late systolic				
Left atrial type A	Presystolic	Increased late in inspiration or beginning of expiration	,,	Late	Some are stimulated
Left atrial type B	Late systolic		,,	,,	Injection-discharge time > 20 sec in great majority
Left ventricular	Early systolic in majority	Little or no charge	,,	,,	
Right ventricular	This terminates before mid-systole	,,	Indefinite	Indefinite	All stimulated. Injection-discharge time less than 15 sec in great majority
Aortic pressure receptors	Begins early in systole but volley continues beyond mid-systole	Increased late in inspiration	Reduced or abolished	Late	Not stimulated
Pulmonary arterial baroreceptors (see text)	,,	Increased early in inspiration	?	?	?

then *v* impulses will appear. This may be accompanied by a reduction or absence of the *a* burst which may be due to reduced amplitude of the *a* wave, reduced systolic tension or if these are unchanged to the dynamic "off effect" (KATZ 1950). More experiments are clearly needed to establish these points.

C. Ventricular pressure receptors

Electrophysiological studies have led investigators to suspect from time to time the presence of sensory endings in the ventricles. Thus AMANN and SCHAEFER (1943) thought that some of the systolic discharges they noted in fine cardiac branches of the vagus came from endings in the ventricles.

This was doubted by Jarisch and Zotterman (1948) who thought that the discharges from endings in the ventricles were conveyed by the smallest medullated fibres and perhaps also by non-medullated fibres. Looking at their records from multifibre preparations it appears that Jarisch and Zotterman were partly right because the systolic discharge seen by Amann and Schaefer can now be safely attributed to arterial and pulmonary baroreceptors and type B atrial receptors. On the other hand Jarisch and Zotterman felt that the fibres from the ventricles did not have any cardiac rhythm, and that they could be aroused by artificial stimuli such as pinching the ventricles. They could also be aroused by grossly distending the ventricles and Jarisch and Zotterman therefore thought that the endings constituted some sort of defence mechanism. However, it must have been difficult for the authors to be certain since they were recording impulses from a large number of fibres simultaneously. Indeed, if size of the spikes is used as a criterion of identity of a fibre, then records of impulse activity in many fibres will give the impression that there are many more fibres than there actually are, because the random summation of impulses gives the impression that the summated potential is arising from a large-spiked fibre.

A third kind of afferent fibre believed to arise from the endings in the ventricles was reported by Whitteridge (1947, 1948). These, unlike those of Jarisch and Zotterman (1948) had a cardiac rhythm and were characterised by having an early systolic burst of impulses almost immediately after the QRS complex of the ECG coincident with the isometric contraction phase (Whitteridge 1948). Since these endings were excited by some cardiovascular mechanical event, Whitteridge concluded that these endings could be located in the ventricles because the discharge appeared too early for it to arise from arterial endings and the absence of the *a* burst made their location in the venous side unlikely. Subsequently these fibres were also isolated by Dickinson (1950) and by Pearce (1951). Whitteridge pointed out that these fibres might arise in the ventricle or septum, but that their exact origin had to be determined (Whitteridge 1953).

Subsequent work [Paintal 1955 (b); Heymans and Neil 1958; Neil and Joels 1961] has confirmed the presence of endings in the ventricles as predicted by Whitteridge but those noted by Jarisch and Zotterman (1948) remain to be confirmed. However, it is possible that they are the same endings because it is possible that Jarisch and Zotterman could not discern the early systolic discharge in these fibres in their multifibre preparations, but when they pinched the ventricles or injected veratrine the pronounced discharge in the fibres made the presence of the endings noticeable. Undoubtedly, there are endings in the ventricles which may show no more than an impulse during each cardiac cycle, and the activity of which increases enormously following injection of products of veratrum [see Fig. 6, Paintal

1955 (b)]. These are the ventricular pressure receptors described below (Figs. 23, 24).

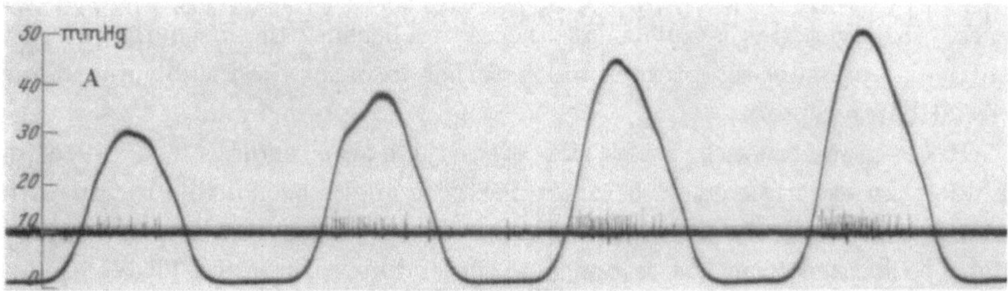

Fig. 23. Impulses in ventricular fibres from a frog's heart showing that increase in ventricular pressure following occlusion of the ductus arteriosus produced increased activity in ventricular receptors (KOLATAT, KRAMER and MÜHL 1957)

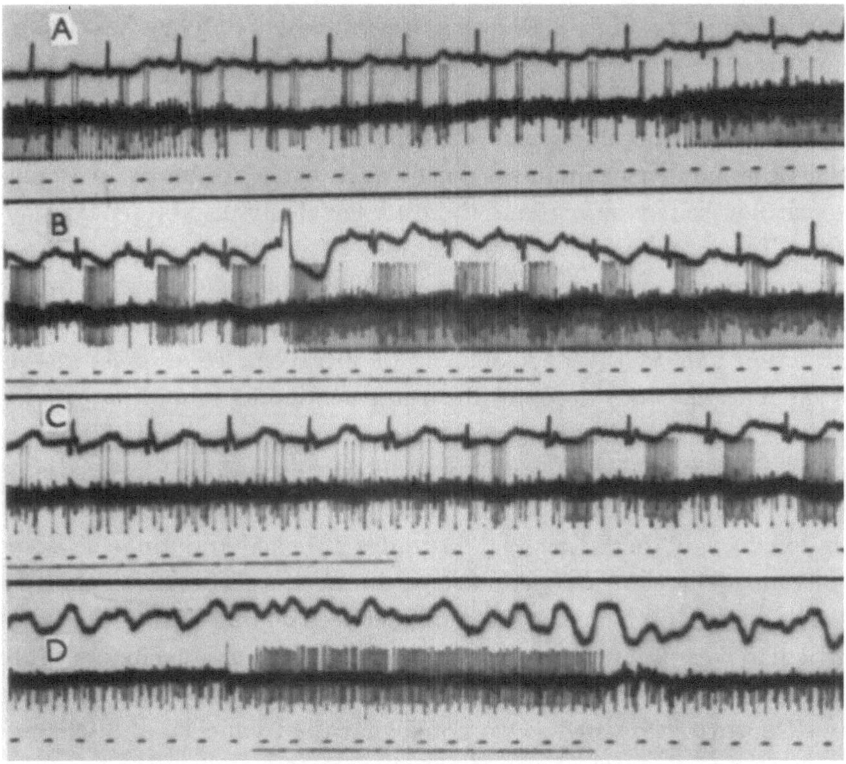

Fig. 24 A—D. Impulses in a right ventricular fibre (large spiked fibre above baseline). A is a normal record with open chest and artificial ventilation. The large spikes below the baseline are from a pulmonary stretch fibre. B and C show respectively the effect of occluding the pulmonary artery and right *a-v* junction, end of signal indicates end of occlusion. In D the right ventricle was pressed during signal after clamping the pulmonary artery and right *a-v* junction. Responses from this fibre suddenly ceased when the ventricle was incised transversely across its middle. From above downwards in each record: e.c.g., impulses in fibres; time in $^1/_{10}$ sec; and signal [PAINTAL 1955 (b)]

Identification of ventricular pressure receptors. These endings are identified by the fact that they possess an early systolic discharge of impulses clearly earlier in onset than the burst of impulses in aortic pressure receptors (Fig. 24).

For purposes of identification it is a useful procedure to have the separated aortic nerve at hand in order to compare simultaneously the discharge in a fibre suspected of arising from the ventricles with that from the whole aortic nerve. However, this is not usually necessary because the discharge in most ventricular pressure receptors is much earlier in onset than that in systemic arterial baroreceptors.

Other characteristics useful for identifying the endings are given in Table 3. In experiments with intact chests it might be possible to say with some luck tentatively in which chamber a particular ventricular receptor might be located from the response of the ending to inflation (Table 3) but to be certain the chest has to be opened and the ending located according to procedures already known [Paintal 1955 (b)]. The most certain information is obtained by isolating the segment of the ventricle in which the ending is located, making sure that the ending is stimulated only by squeezing this segment (Fig. 24 D), a response that is suddenly abolished when this segment is cut away from the rest of the heart [Paintal 1955 (b)].

Responses of ventricular pressure receptors. Under ordinary experimental conditions the number of impulses during each cycle are few when compared to other cardiovascular endings and there are few if any respiratory fluctuations (Table 3). The latency of the first impulse after the Q wave of the ECG is between 20 and 50 msec in most fibres. While this figure is useful it would be more valuable to know the relation of the discharge to the isometric phase of contraction and to the development of intraventricular pressure. These figures do not take into account the total conduction time in the afferent fibres which is most probably of the same order as that in atrial fibres since the conduction velocities of two ventricular fibres determined so far are 14 and 19 m/sec respectively (unpublished observations). So far it has been possible to conclude that some of the endings are excited during the isometric phase of ventricular contraction and some during the early part of the ejection phase. The explanation for this difference between endings is that the threshold of ventricular pressure at which the different endings are excited is probably different. Similar observations have been made in ventricular receptors of frogs by Kolatat et al. (1957) who noted that the frequency of discharge was linearly related to the rate of rise of pressure.

There is also some evidence of adaptation of the endings to pressure because the discharge ceases before peak ventricular pressure is reached. This could also be due to the endings being unloaded during muscular shortening, i.e. during the ejection phase. This would imply that they are distension-sensitive, which is not likely because the endings are not stimulated during the *a* wave of atrial systole, i.e. when ventricular volume is maximum even when the ventricle is grossly distended by impeding the outflow from it (Fig. 24 B).

The endings can clearly signal changes in intraventricular pressure (Figs. 23, 25) because a greater rise in pressure elicits a greater number of impulses and a higher frequency of discharge. Fig. 23 shows that the same is true of endings in the frog's ventricles (KOLATAT et al. 1957). Thus in the fibre shown in Fig. 25 premature ventricular contractions produced fewer impulses and a lower peak frequency of discharge than normal contrac-

tions. The pattern of discharge also seems to follow the ventricular pressure and it does not change even when the pressure is increased considerably (Figs. 23 and 25).

The most striking and characteristic feature of ventricular pressure receptors is that all of them, unlike other cardiovascular mechanoreceptors, are stimulated by relatively small doses of veratrum alkaloids [PAINTAL 1955 (b); NEIL and JOELS 1961], i.e. amounts that are needed to elicit the Bezold-Jarisch effect (see KRAYER and ACHESON 1946; KRAYER 1961). These sub-

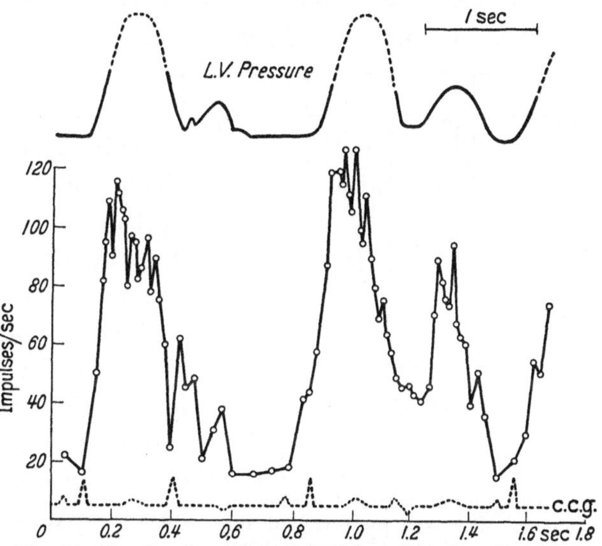

Fig. 25. Graph of response in a left ventricular fibre during occlusion of the aorta. The interrupted portions of the pressure curve indicate the places where the great rise in pressure overshot the pressure registering limits of the manometer, solid part is a reproduction of the original record. The ECG is given below [PAINTAL 1955 (b)]

stances cause a dramatic increase in activity of the endings which sets in after a latency ranging from 5—15 sec depending on the type of alkaloid injected into the right atrium [PAINTAL 1955 (b)]. The duration of stimulation again depends on the nature of the alkaloid and may vary from 1—18 sec in the case of veratridine and 1—15 min after veriloid. The response of these endings to veratridine is so characteristic and constant that it can be used as a means of identification of the ending in cases of difficulty (Table 3). The chief effect of the alkaloids is the production of a continuous discharge which may attain a high frequency and which overshadows the cardiac rhythm. As shown subsequently, such results indicate that the endings are stimulated and simultaneously desensitized by the alkaloids [PAINTAL 1957 (b)].

D. Pulmonary arterial baroreceptors

Although the existence of endings in the region of the pulmonary artery had been suspected for a long time, it was only after SWAN and WHITTERIDGE (1956) had firmly located one ending to the bifurcation of the pulmonary artery that their existence was clearly established electrophysiologically. In this single experiment Swan and Whitteridge were able to obtain the essential information concerning the behaviour of the endings. Thus they showed that activity in the ending ceased abruptly when the artery was occluded and it returned in the first beat after the vessel was released (Fig. 26). Peak frequency of the discharge was attained at 50—65 msec from the beginning of the Q wave of the ECG. From Fig. 26 it can be estimated that the first impulse appeared at about 45—55 msec. Finally they found that the ending could be stimulated by local mechanical stimulation near the bifurcation.

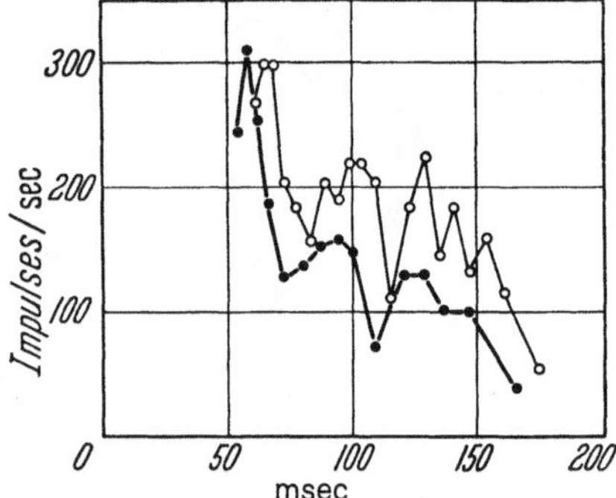

Fig. 26. Pattern of discharge in a pulmonary arterial baroreceptor of a cat. Graph —o—o— shows the discharge that appeared in the first cycle following releasing a loop around the origin of the pulmonary artery. Graph —•—•— shows the discharge pattern 3 or 4 beats later. Time intervals were measured from the Q wave of the ECG. Note that the high frequency pattern of discharge is similar to that of aortic baroreceptors, and is quite unlike the crescendo pattern commonly seen in type B receptors (Fig. 16). (Swan and Whitteridge 1953, unpublished observations)

Subsequently more valuable information about the endings has been obtained in cats by Bianconi and Green [1959 (b)] and in dogs by Coleridge and Kidd (1960, 1961) and Coleridge, Kidd and Sharp (1961).

Identification of pulmonary arterial baroreceptors. There is apparently no published information on the isolation and identification of pulmonary arterial receptors with closed chest as all the reported experiments have been done with open chest. While there is no doubt that to be certain the endings must be located after opening the chest, there is much to be said for being able to identify an ending provisionally with closed chest so that observations with closed chest may be made. Such information is valuable and has been secured in the case of all other cardiovascular receptors. Presumably, there will be respiratory fluctuations as in the case of other endings and as noted in one fibre suspected to arise from the pulmonary artery (see Walsh and Whitteridge 1945 and Swan and Whitteridge 1956). The increase in activity will probably follow the fluctuations in right atrial receptors and precede those in aortic receptors as expected from the results of Pearce and Whitteridge (1951), since the endings are stimulated by rise in pulmonary arterial pressure which is their natural stimulus.

Not only have these endings to be distinguished from those atrial type B endings with a discharge of impulses slightly earlier in onset and from arterial baroreceptors in which the timing of the discharge in relation to the ECG is similar but also from atrial type A endings because as shown by Coleridge and Kidd (1960) some pulmonary arterial baroreceptors, especially those located in the right branch of the pulmonary artery may have a more prominent *a* burst of impulses than a systolic burst. Perhaps, as their results would indicate this might happen when the pulmonary arterial pressure is low giving rise to a weak systolic volley. However, infusion of some fluid can distinguish between a true type A ending and a pulmonary arterial

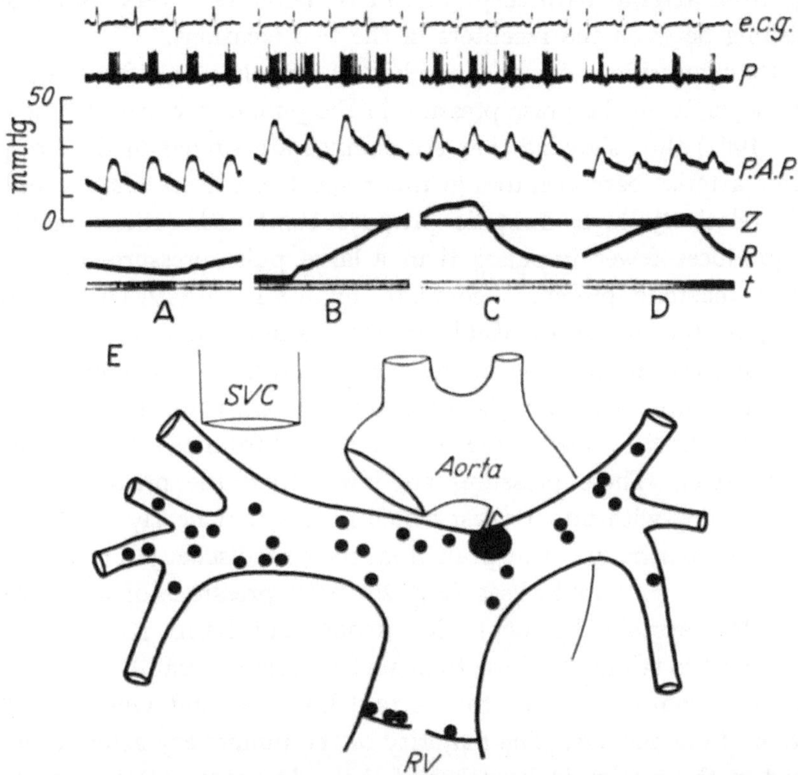

Fig. 27 A—E. Effect of occlusion of lung roots on pulmonary arterial pulse pressure and pulmonary baroreceptor activity. A, before occlusion; B, C and D during occlusion of the lung roots. Intervals of approximately 2,3 and 1 sec between records A and B, B and C, and C and D respectively (COLERIDGE and KIDD 1961). E, the position of 44 pulmonary arterial baroreceptors from which action potentials had been recorded and whose location had been determined accurately by punctate stimulation after the animals had been killed and the pulmonary artery dissected. The position of 33 receptors is indicated by closed circles; the remaining 11 receptors were situated in the solid black area near the attachment of the ligamentum arteriosum. The ascending aorta and the superior vena cava (*SVC*) near its junction with the right atrium which lie anterior to the right pulmonary artery have been omitted for clarity (COLERIDGE, KIDD and SHARP 1961)

receptor because in the latter, the *a* burst is abolished by such an infusion (COLERIDGE and KIDD 1960). COLERIDGE and KIDD believe that the atrial systolic burst is due to distortion caused by atrial contractions. These results therefore show that to be sure the endings must be located after opening the chest.

Natural stimulus. It is now certain that the natural stimulus for these endings is a rise in pulmonary arterial pressure [SWAN and WHITTERIDGE 1956; BIANCONI and GREEN 1959 (a); COLERIDGE and KIDD 1960, 1961]. Thus the discharge is abolished when the pulmonary artery is occluded near the conus, and it is increased when branches of the artery are occluded distal to the location of the ending (Fig. 27B) (Coleridge and Kidd 1960, 1961). A less dramatic increase in discharge is obtained by occluding the aorta [Bianconi and Green 1959 (b)], a procedure that reduces or abolishes the discharge from arterial baroreceptors and is therefore a very useful one for distinguishing between the receptors in the two locations.

More than anything else the number of impulses and the frequency of discharge depends on the pulse pressure in the pulmonary artery (Coleridge and Kidd 1961) thus showing that the endings are similar in their responses to systemic arterial baroreceptors in this respect (see Ead, Green and Neil 1952). A relatively high diastolic pressure combined with a small pulse pressure produces fewer impulses than a large pulse pressure superimposed on a lower diastolic pressure (compare Fig. 27A with 27D). Artificially produced pulsatile pressures yield similar results (Coleridge and Kidd 1961) although the information in this connection is not as complete as that on carotid pressure receptors [Bronk and Stella 1932, 1935; Landgren 1952 (a), (b)]. The threshold of the endings ranged from 15—25/7—13 mm Hg (systolic/diastolic). These pressures are lower than the pressures normally present in the dog with intact chest which means that nearly all the endings must be active normally. The peak frequency of discharge attained in the endings is 200—300 impulses/sec (Fig. 26) with pressures of about 45—50/ 20—25 mm Hg (systolic/diastolic) (Coleridge and Kidd 1961). A smaller force is required to stimulate these than aortic endings because the pulmonary artery is more compliant than the aorta [Bianconi and Green 1959 (b)].

Location of the endings. The majority of the pulmonary arterial receptors are located in the two main branches of the pulmonary artery and near the bifurcation [Bianconi and Green 1959 (b); Coleridge and Kidd 1960, 1961; Coleridge, Kidd and Sharp 1961]. In the case of the dog, Coleridge et al. (1961) have gone a step further and defined more precisely the positions of the endings by means of punctate stimulation with a probe and careful physical isolation of the ending. The positions of the endings located in this way in the dog are shown in Fig. 27E. A notable feature is that there are very few endings in the main trunk of the pulmonary artery. The presence of the endings has also been confirmed histologically [Bianconi and Green 1959 (a); Coleridge, Kidd and Sharp 1961] but it is difficult to be certain that the endings isolated histologically are in fact the ones from which impulses were recorded. Perhaps one could do the experiment after chronic sympathectomy, chronic vagotomy below the nodose ganglion on one side, and above the ganglion on the other, then isolate the ending histologically after studying it electrophysiologically. Much uncertainty about the source of the fibres would be eliminated in this way although the procedure will be rather tiresome.

BIANCONI and GREEN [1959 (a)] have shown that in the cat a large number of fibres from the bifurcation of the pulmonary artery and from the right pulmonary artery run in a small branch that joins the vagus about 2 cm rostral to the azygos vein. In the dog the afferent fibres from the right pulmonary artery run in both vagi but those from the left run in the left vagus (COLERIDGE, KIDD and SHARP 1961). Presumably this is also true in the cat. So far no impulses have been recorded from endings in the ductus arteriosus which is known to possess sensory endings (TAKINO and WATANABE 1937; BOYD 1941; NONIDEZ 1941).

A suitable technique for tracing the nervous pathway is temperature block which COLERIDGE, KIDD and SHARP (1961) used to determine in which branches of the several cardiac nerves the afferent fibres from pulmonary arterial baroreceptors travelled. They found that half the afferent fibres travelled in the recurrent cardiac (which joins the right vagus) and the ventromedial cervical cardiac (which joins the left vagus) nerves.

XI. Gastro-intestinal afferent fibres

Sensory impulses in gastric branches of the splanchnic nerve have been recorded by DELOV (see BYKOV 1958). Impulses from endings in the intestines were recorded by TOWER (1933) in the sympathetic rami of the frog. In the cat this was achieved by GERNANDT and ZOTTERMAN (1946), BROWN and GRAY (1948), MEIER and BEIN (1950) and BEIN and MEIER (1951) by recording from the mesenteric nerves. Unfortunately most of the information is derived from branches containing many fibres and the resulting records are difficult to analyse.

Vagal endings in the gastro-intestinal tract were found accidentally [PAINTAL 1953 (d)] while looking for thoracic receptors responsible for the amidine reflexes following intravenous injections of phenyl diguanide (see DAWES and MOTT 1950; DAWES et al. 1951). Initially it was found that phenyl diguanide [PAINTAL 1953 (b)] and 5-HT (MOTT and PAINTAL 1953) stimulated certain endings that were believed to be located in the lungs. However, further experiments showed that impulses in such filaments could also be produced by intra-aortic injections of phenyl diguanide so that it was clear that the endings were located in the abdominal viscera and not in the lungs. Eventually, the first series of endings encountered were found to be located in the stomach and intestines [PAINTAL 1953 (d), 1954 (a)]. The natural stimulus of those of the stomach was found to be gastric distension (Figs. 28 and 29) and these were therefore termed gastric stretch receptors [PAINTAL 1953 (d)].

As mentioned on p. 96 at least 4 kinds of endings in the gastrointestinal tract are now known, viz. gastric stretch receptors, gastric mucosal chemoreceptors, distension-sensitive intestinal tension receptors and mucosal mechanoreceptors (distension-insensitive). The natural stimuli of each of these is different but they all have one common property, namely that most of the endings are connected to non-medullated afferent fibres (IGGO 1958). The conduction velocities of the fibres are shown in Fig. 7 from which it is clear that it is less than 1.5 m/sec in most of them i.e. they are in the non-medullated range. However, there are fibres with higher conduction velocities

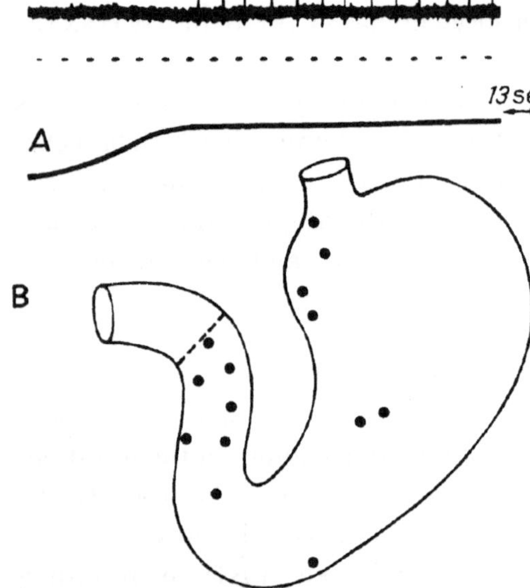

Fig. 28 A and B. A, response of a gastric stretch afferent fibre to a maintained distension of the stomach. From above downwards; impulses in a gastric stretch fibre; time in 0.1 sec; intragastric pressure. The frequency of discharge is unaltered throughout the period of distension [PAINTAL 1953 (d)]. B, location of gastric stretch receptors in the stomach [PAINTAL 1954 (b)]

[PAINTAL 1954 (b)] especially those arising from gastric mucosal receptors (Fig. 7A) and from tension receptors at the lower end of the oesophagus (IGGO 1958).

A. Gastric stretch receptors

These are the endings in the stomach whose chief stimulus is passive distension of the stomach (Fig. 28A) [PAINTAL 1953 (d), 1954 (b)]. They are synonymous with those which IGGO refers to as gastric tension receptors [IGGO 1955, 1957 (a)] and gastric "in series" tension receptors (IGGO 1958).

Isolation and identification. Although these endings were originally isolated by intravascular injections of phenyl diguanide [PAINTAL 1953 (d), 1954 (a)], there is now no need to use this procedure because the fibres can be isolated simply by noting the effect of passive distension of the stomach with a balloon. Thus if a strand showing little activity normally, acquires a prominent discharge of impulses on distension of the stomach it is almost certain to arise from gastric stretch receptors, but in any case pulmonary stretch receptors and gastric mucosal receptors [IGGO 1957 (b)] stimulated by over-distension of the stomach must be excluded. As a rule pulmonary fibres stimulated by gastric distension can be distinguished easily by noting the effect of inflation of the lungs which yields a pronounced discharge in them, but occasionally the excitation is weak and then one is faced with a decision whether the ending is an atypical pulmonary receptor or a gastric stretch receptor stimulated indirectly by inflation of the lungs [see PAINTAL 1954 (b)]. There are two ways of deciding the issue. One is to see if an intra-aortic injection of phenyl diguanide stimulates the ending; if it does so with characteristic latency it is definitely a gastric receptor [PAINTAL 1954 (c)]. If it does not, it is either a gastric stretch receptor not stimulated by phenyl diguanide because there are some that are not [PAINTAL 1954 (c); IGGO 1957 (a)], or it is a pulmonary receptor. In such cases the second method can be used. This consists of distending a balloon placed between the stomach and diaphragm. This will stimulate pulmonary endings in the same way as distending an intragastric balloon, but it usually does not stimulate the gastric endings [PAINTAL 1954 (b)]. Finally, to be certain, the ending can be located by means of mechanical stimuli such as pinching the stomach locally [PAINTAL 1954 (b); IGGO 1955, 1957 (a)].

"Spontaneous discharges". For want of a better term, the term "spontaneous" discharge has been used here and in the rest of this review to cover the discharge of impulses appearing in afferent fibres in the absence of any external stimulation and at present not clearly attributable to a particular natural stimulus. No doubt, the natural stimulus will be found in the future.

Typically, there is no activity in most gastric stretch receptors when the stomach is empty by which is meant that there is no food in the stomach and it is not distended by a balloon (condom) which is invariably inserted into the stomach in such experiments; the volume of the deflated condom is not more than 10 ml. In addition to the condom a small quantity of gastric juice is always present [PAINTAL 1954 (b)]. It is significant, that even under such conditions, the majority of gastric stretch receptors show little or no activity [PAINTAL 1954 (b)]. In the few that do, the discharge is usually unimpressive consisting mostly of irregularly occurring impulses of variable frequency. In about half the fibres that show "spontaneous" discharges, the frequency varies between 0 and about 8 impulses/sec. In other fibres there may be groups of impulses with peak frequencies of about 20—35 impulses/sec with anything from 6—90 impulses in each group. The frequency of the groups is also variable and they appear in the absence of any change in intragastric pressure [PAINTAL 1954 (b)]. This means that if these grouped discharges are due to local contractions of the stomach, the contractions are so weak that they cause practically no rise in intragastric pressure (cf. IGGO 1955).

Natural stimulus. There is now no doubt that the natural stimulus for these endings is distension of the stomach (Fig. 28A) [PAINTAL 1954 (b); IGGO 1957 (a)]. The distension is effective whether it is achieved by inflating a balloon with air or water [PAINTAL 1954 (b); IGGO 1955] or by introducing saline into the stomach [IGGO 1957 (a)]. As is to be expected, the threshold is different for different endings and it varies from 5—150 ml, being less than 50 ml in the majority [PAINTAL 1954 (b); a similar conclusion may be drawn from IGGO's results [IGGO 1957 (a)]. Since the threshold may be expected to vary with the size of the stomach, it is not surprising to find that the threshold in the goat is much higher than in the cat, being of the order of 300—400 ml (IGGO 1955). The threshold also varies with the experimental conditions and attachments to the stomach, but with care it remains constant for long periods [PAINTAL 1954 (b)].

With maintained distension, the discharge adapts slowly in the majority of the endings (Figs. 28 A, 29 A). In some no tendency to adaptation was observed even with rapid and large distensions (Fig. 28 A). In a few fibres the discharge adapted rapidly. It is possible that this might be related to the slow relaxation of the smooth muscle as suggested by IGGO in the case of distension-sensitive intestinal receptors [IGGO 1957 (a)].

Gradual distension of the stomach gives rise to a gradually increasing discharge of impulses [PAINTAL 1954 (b); IGGO 1957 (a)] thus showing that the slow entry of food or water will stimulate these endings in the same way. As shown in Fig. 29 B the introduction of saline at the average rate of about 3 ml/sec was adequate to produce an impressive discharge of impulses.

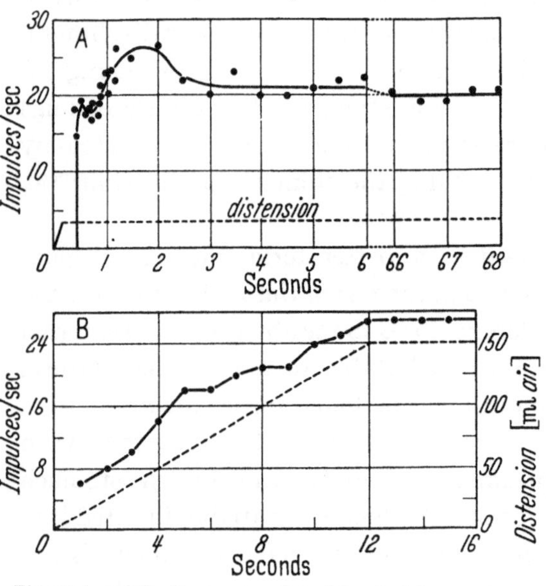

Fig. 29 A and B. Responses of gastric stretch receptors. A shows the response of a slowly adapting gastric stretch receptor to distension of the stomach which was maintained for over 1 min. Distension is indicated by the interrupted line. B, frequency of discharge in a gastric afferent fibre with gradual distension of the stomach. The gastric balloon was distended with 150 ml air in 12 sec as shown by the interrupted line [PAINTAL 1954(b)]

The interval between the onset of distension and the appearance of the discharge at the recording electrodes near the nodose ganglion varies from 0.1—0.5 sec (Fig. 28 A). Initially, the increased delay was attributed to inertia of the balloon [PAINTAL 1954 (b)] because at that time it was erroneously believed that the majority of the afferent fibres were medullated so that even if they had a conduction velocity of about 6 m/sec, the total conduction time would not amount to more than 50 msec. After IGGO's finding that the majority of the fibres have a conduction velocity less than 1.5 m/sec (IGGO 1958), it has become clear that total conduction time will be of the order of 0.2—0.5 sec so that most of the delay between distension and the beginning of the discharge can now be explained on this basis.

The frequency of discharge varies with the degree of distension and is frequently linearly related to the degree of distension [PAINTAL 1954 (b)]. A similar sort of relationship is seen in goats (see Fig. 1 in IGGO 1955). The peak frequency of discharge in these fibres has never exceeded 60 impulses/sec and it usually varies between 25 and 60 impulses/sec [PAINTAL 1954 (b)]. The same is true in the goat (IGGO 1955). This contrasts with the much higher frequencies that can be attained in other vagal afferent fibres. As noted by IGGO [1957 (a)] the peak frequency depends on the rate of gastric distension.

Effect of localised contractions. In the earlier investigation [PAINTAL 1954 (b)] it was found that gastric contractions induced by applying a galvanic stimulus anywhere on the distended stomach did not stimulate the endings although there was a considerable rise in intragastric pressure [PAINTAL 1954 (b)]. On the other hand manual compression of the distended stomach associated with rise in intragastric pressure stimulated the endings, which was attributed to stretching of the stomach at the site of the ending. It was therefore concluded that a rise in intragastric pressure *per se* does not stimulate the ending if it does not stretch it [PAINTAL 1954 (b)]. However, in the goat, a rise in intragastric pressure produced by reflex contraction of the stomach stimulates the endings markedly (Fig. 30E). In such cases it is not known whether the stimulation is due to the rise in pressure causing the ending to be stretched or to contraction of the stomach in the region of the ending. As pointed out by JAMES (1957) it is possible that a gastric contraction at one point might distend the stomach at another containing the ending.

On the other hand it is definite from IGGO's work that local contractions can stimulate the endings markedly (Fig. 30) [IGGO 1955, 1957 (a)] and this may happen without any change in intragastric pressure. Presumably, as believed by IGGO, this may be due to increase in tension in the wall of the stomach, in which case the endings might be considered to be in series with the muscle elements. There seem to be some differences between the endings at the cardiac end from those at the pyloric end because the former could not be stimulated by local contractions (IGGO 1955). Some endings cannot be stimulated when the stomach is empty and some require a certain threshold of distension before superimposed contraction can stimulate them (Figs. 30A, B and C) (IGGO 1955).

Location of the endings. Most of the endings are localised in the pyloric portion of the stomach, some at the cardiac end, especially near the lesser curvature (Fig. 28B) [PAINTAL 1954 (b); IGGO 1955]. It was suggested [PAINTAL 1954 (b)] that the endings were probably located in the smooth muscle itself because stroking the mucous membrane or the serosal surface of the stomach did not stimulate the endings. One can now be more certain about this because IGGO [1957 (a)] has shown that the receptors are unaffected by removal of the mucosa and submucosa of the stomach.

Function of gastric stretch receptors. Whether these endings are ultimately found to lie in parallel or in series with muscle fibres it is at least certain that they are stimulated by distension of the stomach, this being the chief stimulus that activates them and by which they are identified [Paintal 1954 (b); Igo 1955, 1957 (a)]. In addition the endings may signal certain types of

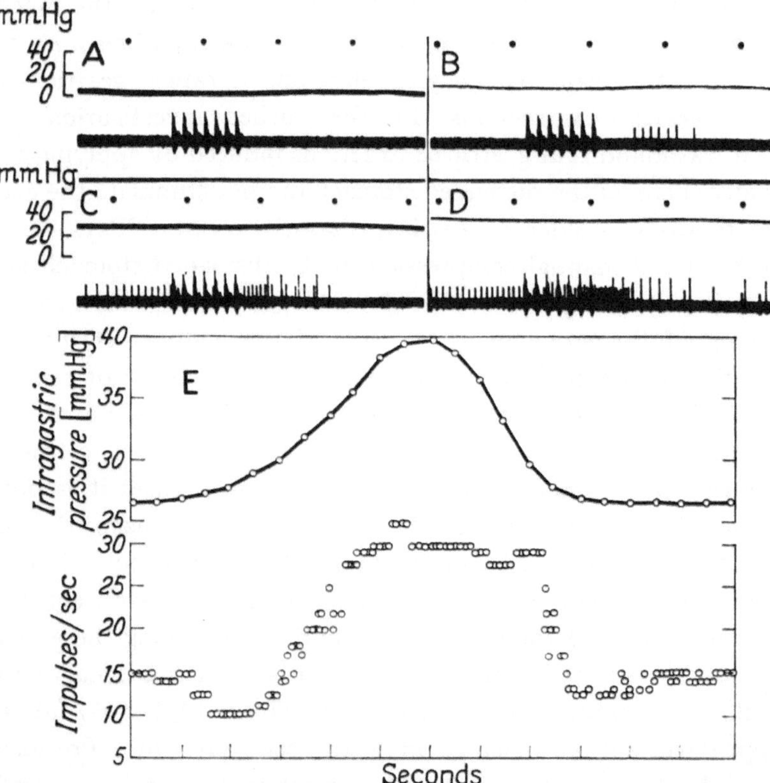

Fig. 30 A—E. The effect of gastric distension and of contractions, produced by stimulating gastric motor fibres in the cervical vagus, on the discharge of impulses in gastric afferent fibres of goats. The volume of inflation was 100 ml in A, 200 ml in B, 300 ml in C and 400 ml in D. The stimulus artifacts can be seen alone in A, in which no impulses were recorded. At the higher volumes of inflation a second fibre was active during the contractions. Time 1 sec. E, the discharge of impulses in a slowly adapting fibre during sustained gastric distension of 1200 ml and a reflex contraction elicited by the distension. Above: intragastric pressure; below: frequency of discharge of impulses in the single fibre (Iggo 1955)

gastric contractions causing the endings in the pylorus to be stimulated (Iggo 1955). These contractions are more effective when the stomach is distended (Figs. 30 B, C, D) than when it is empty (Fig. 30 A) (Iggo 1955). In fact the majority of the fibres are silent normally with an empty stomach, this being the chief reason why the endings had not been encountered earlier [Paintal 1954 (b)].

Therefore during intake of food or water activity in these fibres will increase and rise to a peak when the stomach is full. Thereafter, there might be some reduction in the discharge because the endings adapt slowly, but with the slow rate of ingestion of food this will be insignificant. However, the total afferent inflow will not change owing to the increase in the discharge by the superimposed gastric contractions as noted by IGGO (Fig. 30). The discharge will fall when the stomach begins to empty and most of the fibres will become silent when the stomach is fully emptied although about a third of them will continue to discharge impulses at about 1—10 impulses/sec [PAINTAL 1954 (b)]. Some of these impulses may be related to contractions of the empty stomach but this needs confirmation.

This description of how the discharge is influenced by gastric contents and gastric contractions reconciles certain superficial differences in the conclusions advanced by IGGO and myself respectively (see IGGO 1955). In this connection although it might appear that there is a basic difference in the terms gastric stretch receptors and gastric tension receptors, it matters little which term is used because most endings that are stimulated by increase in tension produced by passive stretch or by contraction of muscular elements are commonly referred to as stretch receptors, e.g. GOLGI tendon organs known to be in series with the muscle fibres (HUNT and KUFFLER 1951).

No confusion need arise from IGGO's description of one gastric "stretch" receptor [IGGO 1957 (b)] if it is realised that according to criteria of identification used such endings constitute typical gastric stretch receptors as described above. The fact that an ending is not excited by gastric contraction but is stimulated by small distensions of the stomach should not make any difference in its identification as a gastric stretch receptor because there are endings that are not stimulated by gastric contractions unless the conditions are favourable (see IGGO 1955), and if the investigator is not persistent enough!

B. Gastric mucosal chemoreceptors

IGGO has described certain endings in the mucosa of the stomach which yield a prominent discharge of impulses when either alkaline or acidic solutions are introduced into the stomach. This discharge which seems to depend on the degree of acidity or alkalinity of the solution (Fig. 31) can be obtained repeatedly in the same endings [IGGO 1957 (b)].

Occassionally some difficulty may arise in identifying these endings from gastric stretch receptors because a few endings (about 15 %) are stimulated by over-distension of the stomach. A further complication is that both groups of endings are stimulated by stroking the mucosa. In fact, the gastric stretch receptors (i.e. IGGO's tension receptors) yield a more prolonged discharge to light pressure on the mucosa while the mucosal endings respond with a brief burst of 1—10 impulses [IGGO 1957 (b)]. In this connection it seems that the receptive field of the mucosal endings may be of the order of about 5 cm².

One characteristic feature of these endings which IGGO has found consistently is that unlike gastric stretch receptors they are not stimulated by local contractions induced by faradic stimulation of the serosa or mucosa.

The mucosal endings are located in all parts of the stomach. By carefully and gradually scraping the mucosa with a scalpel blade, Iggo was able to conclude that the endings were located in the mucosa because removal of only a third of the mucosa was sufficient to abolish the responses from many of the endings. Iggo also concluded that a single nerve fibre might innervate an area perhaps as large as 5 cm² but the evidence in this connection is not

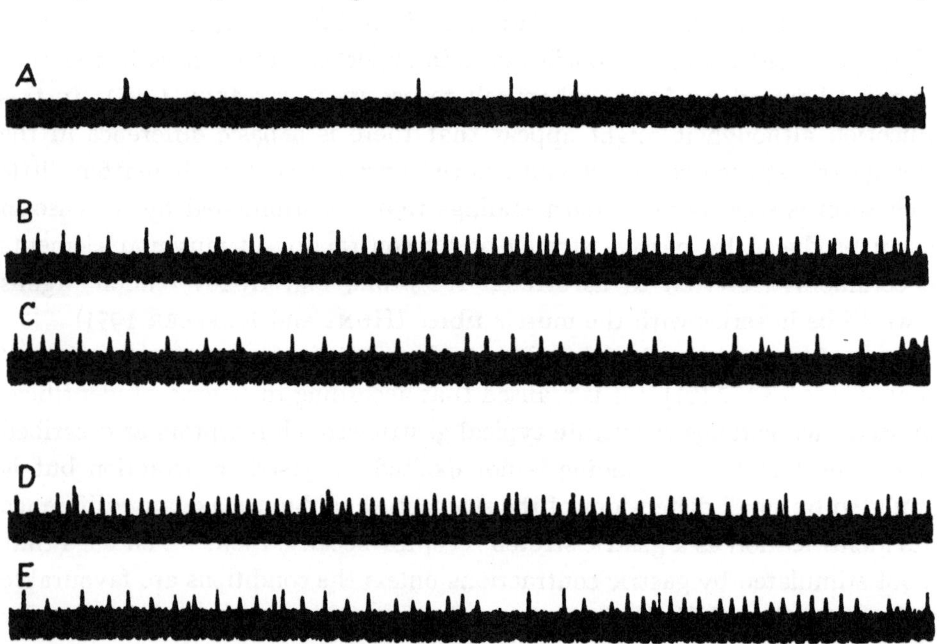

Fig. 31 A—E. The response of an acid-sensitive receptor in the intact stomach to solutions put into the stomach through the cardiac sphincter, A, 0.03 N-HCl, p_H 1.5; B, 0.1 N-HCl; C, the same unit 30 sec later with the acid still in the stomach; D, 0.2 N-HCl; and E 30 sec later as in C. Time marks, 1 sec [Iggo 1957 (b)]

conclusive because the receptive field may be larger than the area actually innervated by a sensory fibre.

The endings are not stimulated by distilled water, Ringer-Locke solution, 0.6 M NaCl, peptone, glucose or sucrose solutions, by 70% ethanol, or even by a 30% suspension (*w/v*) of culinary mustard. On the other hand, they are stimulated either by acidic or by alkaline solutions, *never by both*. This specificity is curious because while it is understandable that the acid-sensitive receptors might signal the activity of the gastric contents normally, the alkali-sensitive ones are never likely to be excited [Iggo 1957 (b)].

Since the acid-sensitive ones yield a slowly adapting discharge to the presence of acidic solutions, a response that varies with the degree of acidity (Fig. 31), it may be concluded that they are slowly adapting chemoreceptors and that they function as p_H receptors [Iggo 1957 (b)].

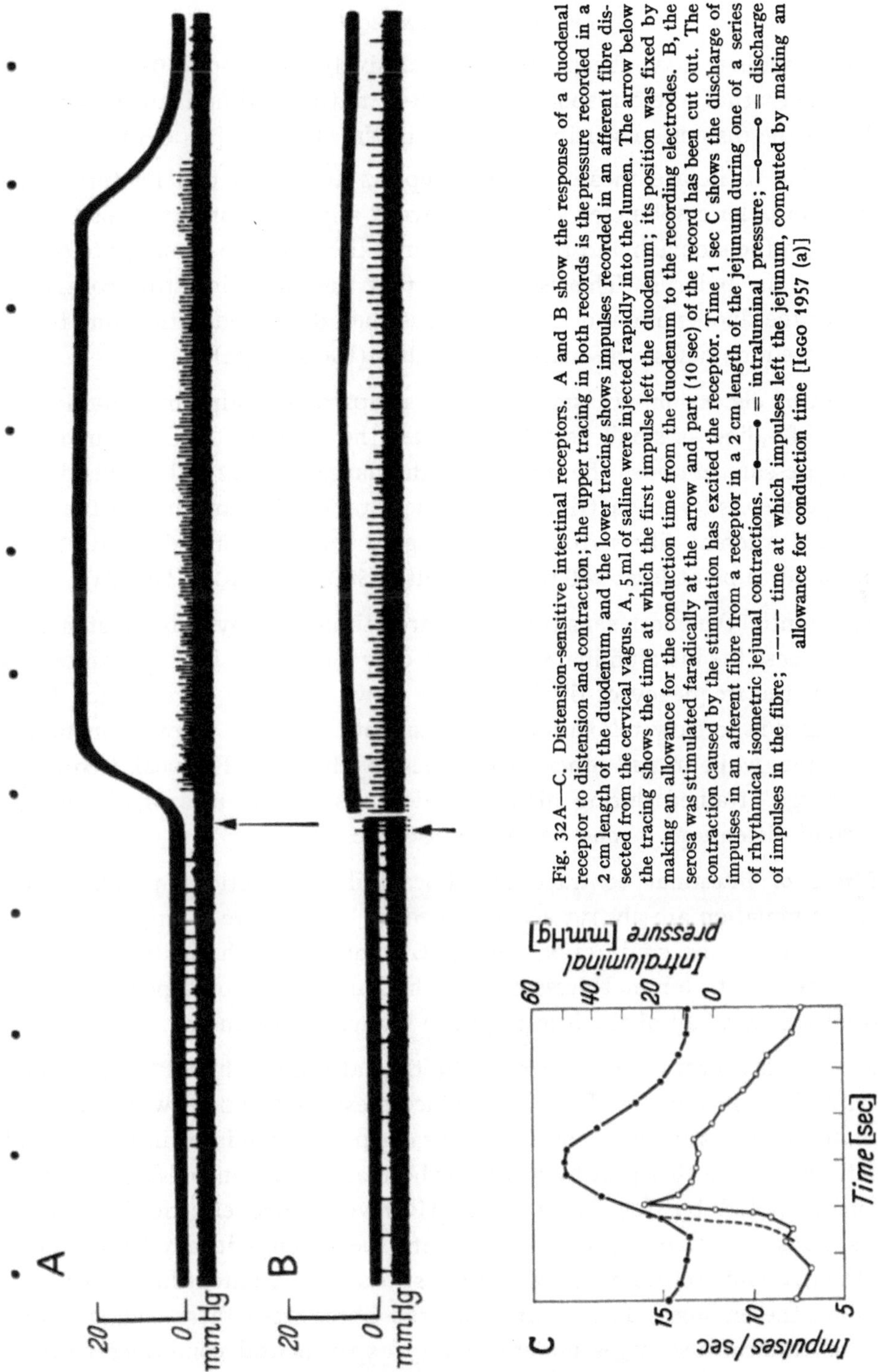

Fig. 32A—C. Distension-sensitive intestinal receptors. A and B show the response of a duodenal receptor to distension and contraction; the upper tracing in both records is the pressure recorded in a 2 cm length of the duodenum, and the lower tracing shows impulses recorded in an afferent fibre dissected from the cervical vagus. A, 5 ml of saline were injected rapidly into the lumen. The arrow below the tracing shows the time at which the first impulse left the duodenum; its position was fixed by making an allowance for the conduction time from the duodenum to the recording electrodes. B, the serosa was stimulated faradically at the arrow and part (10 sec) of the record has been cut out. The contraction caused by the stimulation has excited the receptor. Time 1 sec C shows the discharge of impulses in an afferent fibre from a receptor in a 2 cm length of the jejunum during one of a series of rhythmical isometric jejunal contractions. ⎯●⎯ = intraluminal pressure; ⎯○⎯○ = discharge of impulses left the jejunum, computed by making an allowance for conduction time [Iggo 1957 (a)]

C. Intestinal mechanoreceptors

There are two kinds of intestinal mechanoreceptors, the tension receptors which are distension-sensitive [Iggo 1957 (a)] and the distension-insensitive ones which are referred to as mucosal mechanoreceptors [Paintal 1957 (c)].

1. Intestinal tension receptors

These endings have been studied exclusively by Iggo [1957 (a), 1958], who believes that their responses and their natural stimuli are in general similar to the distension-sensitive endings in the stomach [Iggo 1957 (a)].

In order to isolate these tension receptors Iggo ties two ligatures, one around the pyloric sphincter so as to prevent any communication of the intestine with the stomach and the other around the intestine about 30 to 60 cm caudad to the pylorus. This isolated part of the small intestine consisting largely of the duodenum and the jejunum is then distended with saline to see if this procedure yields a discharge of impulses (Iggo 1957a).

Effect of distension. In small isolated strips of intestine introduction of volumes of saline as small as 0.2 ml excites the endings. The rise in intraluminal pressure precedes the increase in discharge (Fig. 32A) but this difference in latency is largely attributable to the total conduction time in the non-medullated afferent fibres from the ending to the recording electrodes; this is estimated to be at least 200 msec and often 500 msec [Iggo 1957 (a)].

Apparently the most sensitive units are stimulated by small distensions which increase the intraluminal pressure by only 2 mm Hg. During maintained distension the frequency of discharge falls slowly in some fibres which Iggo believes is due to slow relaxation of the smooth muscle. In other instances it disappears while in still others it becomes rhythmical. The peak frequency of discharge attained during distension depends on the rate of distension; with rapid distensions it may be 35 impulses/sec.

Effect of intestinal contractions. Localised contractions produced by faradic stimulation are always accompanied by a discharge of impulses which lasts as long as the contraction (Fig. 32B), and although this is associated with a rise in intraluminal pressure, the discharge does not depend on it, as contractions in a flap of the intestines can stimulate the ending.

Rhythmical contractions generated by the natural rhythm of the intestines also stimulate the endings (Fig. 32C). Since these contractions were observed in isolated jejunal segments, contraction of one part of the intestines must lead to distension of another part by virtue of the marked rise in pressure of about 50 mm Hg noted by Iggo (Fig. 32C). However, since electrically induced contractions in a flap of intestine stimulate the endings it may be presumed that the rhythmic contractions of the intestines with a patent lumen will also stimulate the endings. This means that the fibres from these endings must show rhythmical discharges under ordinary experimental conditions perhaps of the sort present in fibres from mucosal mechanoreceptors (Fig. 33D).

Location. As in the case of the distension-sensitive receptors of the stomach, those of the intestines are apparently also located in the muscular layers of the intestines becauseresponses from the endings can still be obtained after scraping off the mucosa and submucosa.

2. *Intestinal mucosal mechanoreceptors*

These endings are quite different from the distension-sensitive intestinal receptors because they are not stimulated by distension of the intestines. So far there has been only one report concerning these endings [PAINTAL 1957 (c)].

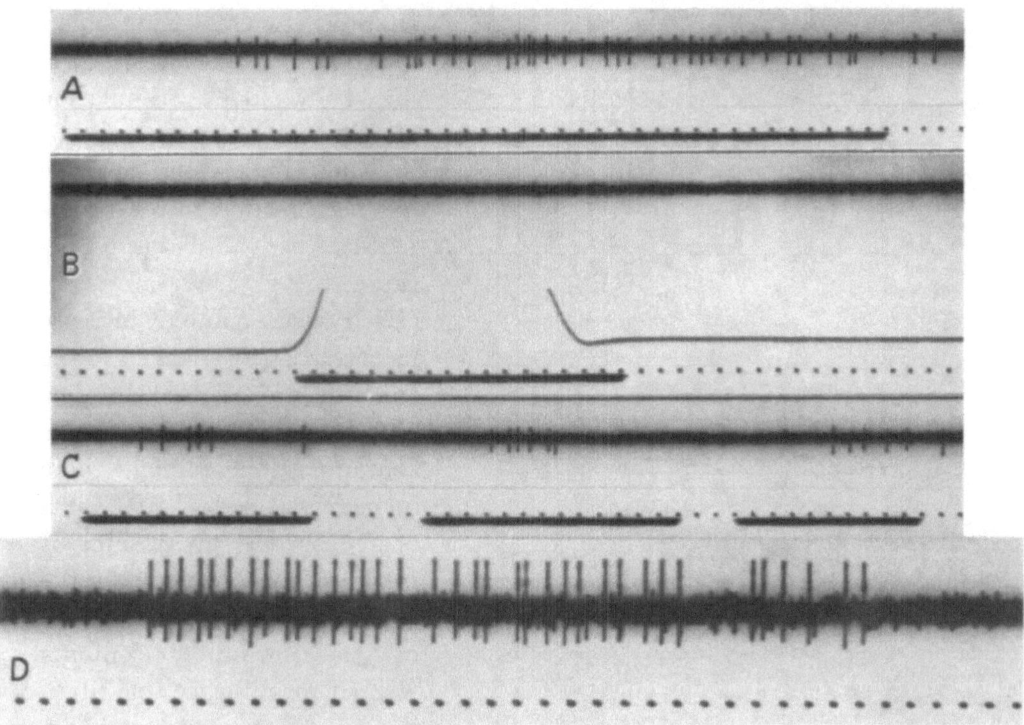

Fig. 33 A—D. Responses of intestinal mucosal mechanoreceptors. A, B and C show the effects of mechanical stimuli. A, the effects on the receptor of squeezing the intestine locally; B, distension of the intestine with 0.9 % NaCl; C, stroking the mucosa locally with rubber tubing. The stimuli were applied at the signals shown in the lowest trace in each record. B also has a record of the intra-intestinal pressure. Time marker 0.1 sec. D shows typical spontaneous discharge of impulses in another fibre. These discharges occurred 3.5 times/min. Time marker 0.1 sec [PAINTAL 1957 (c)]

A useful way of isolating the fibres from these endings is to look for the presence of cyclical bursts of activity of the kind shown in Fig. 33 D recurring at about 3—6 cycles/min. Some gastric endings also have similar cyclical bursts (p. 133). However, since some endings may be silent it is useful to detect the presence of a fibre from them by intraaortic injections of phenyl diguanide [see PAINTAL 1957 (c)]. Distinction from the distension-sensitive receptors [IGGO 1957 (a)] described above is easily achieved by isolating the loop of the intestine containing the receptor and distending it with 0.9 % NaCl; as already mentioned the distension-sensitive ones will be stimulated but not the mucosal endings (Fig. 33 B). The endings are located in all parts of the small intestines

"**Spontaneous**" activity. These endings possess characteristic "spontaneous activity" (revealed by taking continuous records for long periods) consisting of cyclical discharges lasting about 1—6 sec in most fibres, the cycles appearing at about 3—6 times a minute. The peak frequency of discharge varies from 7—26 impulses/sec. This spontaneous rhythm is also seen superimposed on discharges produced by added stimulation [PAINTAL 1957 (c)].

The "spontaneous" activity has been related to the occurrence of peristaltic rushes because these rushes are associated with marked increase in activity (Fig. 34 A). If the contents are allowed to flow past the region of the intestine containing the ending, there is increase in activity which is abolished if an opening is made in the oral side of the receptor. In one case the nature of the intestinal contents seemed to influence the intensity of "spontaneous" activity. Thus as shown in Fig. 34 A the passage of a tape worm apparently produced a greater increase in activity than the passage of the viscid intestinal contents. The "spontaneous" activity was uninfluenced by local application of atropine which made the intestine quite atonic locally. On the other hand, strong contractions produced by local application of acetylcholine did not add to the excitation of the ending (Fig. 34 B). There is a possibility that acetylcholine in some way initiated the activity on some occasions (unpublished observations) but the contraction itself did not seem to be an important factor in influencing the subsequent course of activity (Fig. 34 B). As pointed out by IGGO [1957 (a)] the interpretation of the results is complicated owing to the direct action of acetylcholine on sensory endings [DOUGLAS 1954; PAINTAL 1956 (b); GRAY and DIAMOND 1957; WITZLEB 1959]. In any case it can be

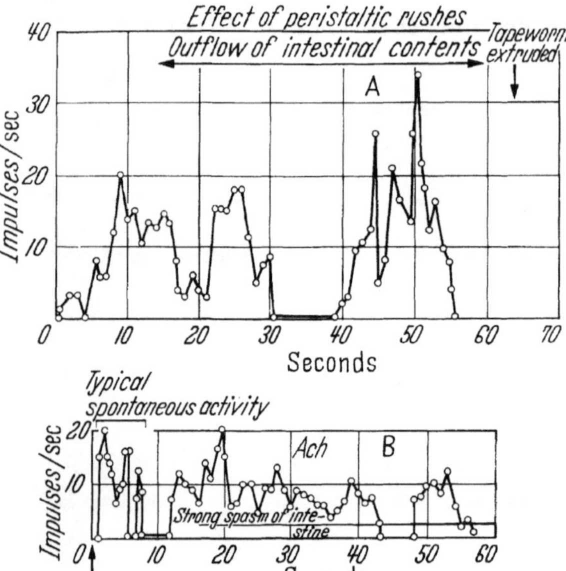

Fig. 34 A and B. Responses of intestinal mucosal mechanoreceptors. A, shows the "spontaneous" activity in a fibre after the end of a period of stimulation by 30% NaCl solution. The intestine had been incised a short distance caudal to the receptor and the appearance of intestinal contents from this opening was observed and is indicated by the horizontal line above. Note the appreciable increase in the frequency of impulses a short while before the tape-worm was extruded, ↓. B, shows that strong spasm of the intestine (indicated by horizontal line) produced by local application of 10% ACh at arrow had little effect on the activity of another receptor; note the spontaneous activity at the beginning of the graph [PAINTAL 1957 (c)]

concluded that contraction of the external intestinal muscles is not the primary factor in exciting the endings.

Responses to mechanical stimuli. Although local distension of the intestine producing a rise in intraluminal pressure of the order of 30—40 mm Hg does not stimulate the endings (Fig. 33 B), it may secondarily set off a train of impulses which is not reduced by lowering the pressure to zero. This discharge is therefore quite different from that seen in distension-sensitive receptors in which the increase in discharge begins and ends with distension (Fig. 32 A). Such secondary discharges have also been seen while squeezing the intestine locally. In some instances this was clear because the local mechanical stimulus yielded a few impulses simultaneously with the stimulus and this was followed by a much longer train of impulses generated by a secondary event initiated by squeezing the muscle.

Injection of intestinal contents such that they flowed past the location of the endings also seemed to produce a discharge aroused by some secondary event following the passage of the contents. The evidence that flow *per se* did not stimulate the endings directly in some cases was clear, although in others it was equally clear that stimulation coincided with the flow of contents. However, in the absence of simultaneous records of flow, it is not possible to be sure, whether in the latter cases, flow was the primary stimulus.

The only mechanical stimulus that seemed to stimulate the endings directly, was stroking the mucous membrane with a solid object such as rubber tubing (Fig. 33 C), but this response could not be obtained in all the endings.

Role of muscularis mucosae. There is no proof but strongly suggestive evidence that the excitation of the endings might depend on the contractions of the muscularis mucosae and this evidence is derived mainly from the effects of introducing strong solutions into the lumen of the intestines [PAINTAL 1957 (b)]. It is known that various mechanical and chemical stimuli cause the muscularis mucosae to contract; it is also known to possess rhythmical activity (BRÜCKE 1851; EXNER 1902; GUNN and UNDERHILL 1914; HAMBLE-TON 1914; KING, ARNOLD and CHURCH 1922; VERZAR and KOKAS 1927; KOKAS 1930; KOKAS and LUDÁNY 1933; WELLS and JOHNSON 1934; KING and ROBINSON 1945; KING, GLASS and TOWNSEND 1947). The mucosal endings are strongly stimulated by 30 % NaCl solution and not at all by a host of other substances in nearly equal concentrations such as magnesium sulphate and sodium sulphate. Similarly only strong NaCl solutions caused the muscularis mucosae to contract as revealed by the characteristic response of the villi. Strong KCl caused a weak contraction and all other substances tried were ineffective.

If the endings are located in the muscularis mucosae, they must lie in a plane that allows the endings to be stimulated by contraction of the muscle but not by stretching the muscle and increasing its tension by distension of the intestine. As suggested [Paintal 1957 (c)], one such plane is that of the villi. However, the crucial experiment consisting of simultaneous recording of contractions of the mucularis muscosae along with the discharge of impulses has yet to be performed in order to show beyond doubt that the muscularis mucosae determines the excitation of the mucosal receptors.

Natural stimulus. The natural stimulus for the endings appears to be the presence or passage of certain intestinal contents. These may stimulate the endings directly but most of the evidence shows that the excitation is brought about indirectly through contractions of the muscularis mucosae. The stimuli which may be mechanical or chemical, may cause the muscularis mucosae to contract directly, or reflexly through a local reflex initiated by some sensory mechanism in the epithelium. These contractions then stimulate the endings, the duration and intensity of the stimulation depending on the degree and duration of contraction [Paintal 1957 (c)].

XII. Composition of vagal rootlets

Several investigators interested in reflexes mediated through the vagus have cherished the hope that there might be a functional grouping of the sensory rootlets of the vagus to enable selective stimulation or elimination of particular afferent fibres. That this is unlikely will become evident from the following account.

The rootlets of the vagus emerge from the medulla in line with those of the glossopharyngeal and accessory nerves so that it is seen that these rootlets together are arranged in three groups, rostral, middle and caudal corresponding to the glossopharyngeal, vagal and the bulbar rootlets of the accessory nerve (Fig. 35 A). If the middle group is viewed dorsally, it is possible to make out that it consists of about 10 or 12 rootlets (Fig. 35 A) (Cadman 1900; Foley and Du Bois 1934; Paintal 1952; Bonvallet and Sigg 1958; Baumgarten and Aranda 1959; Baumgarten, Koepchen and Aranda 1959). Although the bulbar rootlets of the accessory nerve proceed with its spinal root, difficulty may often be experienced in distinguishing them from the most caudal rootlets of the vagus as shown in Fig. 35 A (Paintal 1952). One reason for this is that like the glossopharyngeal, the accessory leaves the skull with the vagus through the jugular foramen but unlike the glossopharyngeal which has a separate dural sheath, the accessory has a common sheath with the vagus. For this reason, one can often distinguish the glossopharyngeal rootlets clearly from those of the vagus (Fig. 35 A).

Earlier work suggested that the afferent and efferent fibres leave the medulla by separate rootlets (Chase and Ranson 1914; Ranson and Mihalik 1932; Ranson, Foley and Alpert 1933). This was confirmed by Foley and Du Bois (1934), using degeneration techniques which showed that the bulbar rootlets of the accessory nerve are composed entirely of motor fibres and that this is also true of the most caudal rootlets of the vagus while the sensory fibres (nodose bundle) of the vagus may enter the bulb alone or with motor or jugular fibres. In another paper Du Bois and Foley (1936) concluded that the motor rootlets of the vagus along with the bulbar rootlets of the

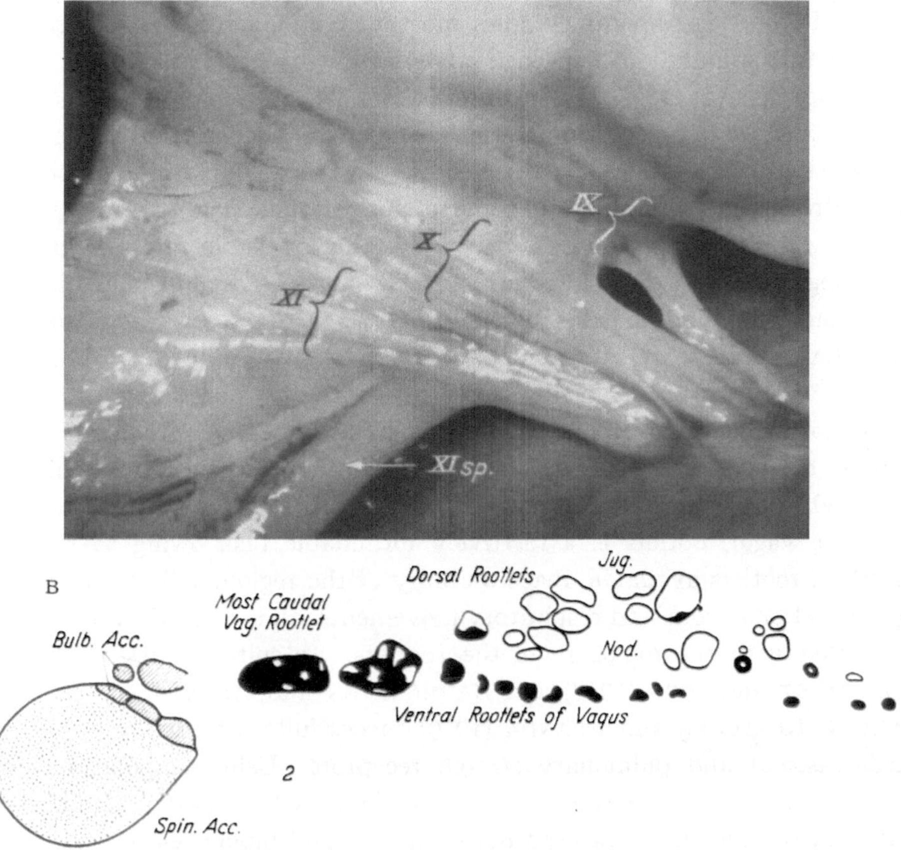

Fig. 35 A and B. Arrangement of the rootlets of the vagus, accessory and glossopharyngeal nerves. A is a photograph of the rootlets as they appear in the pontocerebellar angle (Baumgarten and Aranda 1959). B is a section of the rootlets (vagus and accessory nerves) near the medulla. The sensory portions of the vagus (i.e. undegenerated) are in white. The degenerated vagal rootlets (i.e. motor) are in black (Du Bois and Foley 1936). Note that the most caudal rootlets of the vagus are motor and that the jugular rootlets (cutaneous) form a prominent group in the middle

accessory [which form the recurrent laryngeal nerve (Du Bois and Foley 1936; Lemere 1932)] are assembled in single file ventral to the sensory rootlets of the vagus (Fig. 35 B). This is true only near the medulla because near the jugular ganglion this strictly dorsoventral relation is lost (Du Bois and Foley 1936). It is important to realise that the most caudal rootlets of the vagus are motor (Fig. 35 B), that they cannot be readily differentiated from the bulbar rootlets of the accessory (also motor) and that in practice it is difficult to separate them from the other dorsal sensory rootlets of the vagus although Fig. 35 A suggests that it might not be so difficult. However, this photograph is not really representative of what one sees *during* an experiment because it is difficult to get the area free of blood and there is seldom such a large distance between the medulla and the jugular foramen. As a rule the length of the rootlet available for recording or stimulation is no more than a few millimeters.

The information concerning the functional composition of the rootlets obtained from experimental studies mostly involving study of reflex effects of electrical stimulation combined with section of selected rootlets is quite confusing (Kreidl 1895; Beer and Kreidl 1896; Cadman 1900). This is to be expected because stimulus spread to neighbouring rootlets is almost unavoidable considering that there is barely room enough to insert a pair of electrodes and in some experiments unipolar electrodes were used. In view of this, therefore the only way of obtaining reliable information, is by recording impulses from individual rootlets. This has been achieved in recent years but only with partial success (Paintal 1952; Bonvallet and Sigg 1958; Baumgarten and Aranda 1959; Baumgarten, Koepchen and Aranda 1959).

Sensory impulses in vagal rootlets. Unlike recording impulses in dorsal rootlets of spinal nerves which is even easier than recording from filaments dissected from a nerve (cf. Hunt and Kuffler 1951; Granit 1955), recording from the vagal rootlets is a relatively formidable task owing to the small length of rootlets available, inaccessability of the region, artifacts introduced by arterial pulsations and respiratory movements, and worst of all the bleeding associated with the exposure of the rootlets. In spite of these difficulties, Bonvallet and Sigg (1958) and Baumgarten and Aranda (1959), Baumgarten, Koepchen and Aranda (1959) successfully recorded impulses from cardiovascular and pulmonary stretch receptors. Using unipolar recording electrodes Bonvallet and Sigg recorded impulses from gastric endings also. Baumgarten and Aranda used both unipolar and bipolar electrodes.

Impulses from pulmonary stretch receptors can be recorded from nearly all rootlets of the vagus but they are more frequently encountered in the caudal rootlets (Fig. 36) of the vagus (Paintal 1952; Bonvallet and Sigg 1958; Baumgarten and Aranda 1959; Baumgarten, Koepchen and Aranda 1959). In one instance they could be recorded from a rootlet of the accessory also (Paintal 1952). On the other hand impulses from arterial baroreceptors could be recorded almost exclusively from the rostral group (Bonvallet and Sigg 1958; Baumgarten and Aranda 1959; Baumgarten, Koepchen and Aranda 1959). Activity from atrial type A and type B receptors was recorded from the middle group only (Baumgarten and Aranda 1959; Baumgarten, Koepchen and Aranda 1959).

Baumgarten and Aranda (1959) concluded that although there is a mixture of respiratory and cardiovascular fibres in the middle rootlets, the incidence of mixed rootlets decreases in the rostral and caudal group. While this may be true of mixtures concerning only pulmonary stretch fibres and depressor fibres, it would be surprising to find any rootlet containing only one type of afferent fibres although there may be a superficial predominance of one type electrophysiologically (Fig. 36). This is because there are medullated afferent fibres in the cervical vagus from at least 14 different types of endings (p. 95). To this should be added at least 7 other kinds in the superior laryngeal nerve [ANDREW 1954 (a), (b), 1956 (a), (b), 1957] and about 6 types of cutaneous endings (ADRIAN 1928; ZOTTERMAN 1939; HUNT and McINTYRE 1960) in the auricular nerve. Ignoring the non-medullated fibres, this makes a total of about 27 types of discharges that could be recorded from the rootlets. As

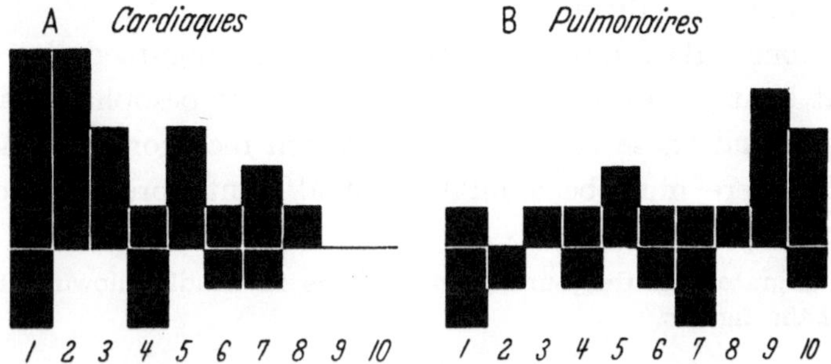

Fig. 36. Frequency of occurrence of cardiovascular and respiratory discharges in the rootlets of the vagus numbered from 1 to 10. Rootlet number 1 is the most rostral and number 10 the most caudal vagal rootlet. The upper histograms represent discharges in dorsal rootlets and the lower ones in ventral rootlets. A, cardiovascular discharges; B, discharges from respiratory mechanoreceptors (BONVALLET and SIGG 1958)

shown in Fig. 35 the fibres of the auricular nerve with their cell stations in the jugular ganglion tend to run in the middle group of rootlets. This fact has been ignored.

Even if one considers only the medullated afferent fibres simple arithmetic will show why it is hardly possible for any rootlet to contain afferent fibres from one type of ending. There are about 7000 medullated afferent fibres running in the vagal rootlets if one adds up all the afferent fibres in the cervical vagus, superior laryngeal, aortic and auricular nerves. If one assumes that there are about 10 or 12 sensory rootlets (PAINTAL 1952; BONVALLET and SIGG 1957; BAUMGARTEN and ARANDA 1959) (Fig. 35) then there should be on an average about 600 to 700 medullated afferent fibres in each rootlet. Even if one can make out activity in 10 fibres in any one rootlet, one can have no idea what is happening (or will happen on proper stimulation!) in the remaining 600 odd fibres. Again consider the fibres from the arterial baroreceptors which run mainly in the aortic nerve. There are only about 300 such medullated fibres (AGOSTONI et al. 1957) (Fig. 2) and even if all of them were to travel in one rootlet, there must still be at least half as many fibres from other endings in the same rootlet. A mixture of cardiovascular afferent fibres with certain others which are less active and are overshadowed by the cardiovascular rhythm even in the most rostral rootlets is therefore inevitable, especially since there are about 3 or 4 rootlets in the rostral group (i.e. altogether about 2 to 3 thousand medullated afferent fibres). Actually, the data of Baumgarten and Aranda (1959) and Baumgarten, Koepchen and Aranda (1959) shows that this must be so because only 17 % of their 36 filaments had a pure cardiac rhythm, and the remaining 83 % a mixture of respiratory and cardiac rhythms. Fig. 36 illustrates this mixture clearly. It must be pointed out that these investigations have ignored the fibres from endings which are not active under ordinary experimental conditions such as chemoreceptors, tracheobronchial receptors, cutaneous receptors and others, like these, which need special procedures to stimulate them.

Finally if one takes into account the 27000 non-medullated afferent fibres from at least 5 known kinds of endings in the oesophagus and gastro-intestinal tract, and those arising from deflation receptors etc. there can be no doubt that there must be a mixture of afferent fibres in every rootlet of the vagus.

The author is grateful to the journals and authors for kindly allowing the reproduction of some of the figures.

References

Adrian, E. D.: The impulses produced by sensory nerve endings. Part I. J. Physiol. (Lond.) 61, 49—72 (1926).
— The basis of sensation. The action of sense organs. London: Christophers 1928.
— Afferent impulses in the vagus and their effect on respiration. J. Physiol. (Lond.) 79, 332—358 (1933).
—, and D. W. Bronk: The discharge of impulses in motor nerve fibres. Part I. Impulses in single fibres of the phrenic nerve. J. Physiol. (Lond.) 66, 81—101 (1928).
—, and Y. Zotterman: The impulses produced by sensory nerve endings. Part 2. The response of a single end-organ. J. Physiol. (Lond.) 61, 151—171 (1926).
Agostoni, E., J. E. Chinnock, M. DeBurgh Daly and J. G. Murray: Functional and histological studies on the vagus and its branches to the heart, lungs and abdominal viscera in the cat. J. Physiol. (Lond.) 135, 182—205 (1957).
Amann, A., u. H. Schaefer: Über sensible Impulse im Herznerven. Pflügers Arch. ges. Physiol. 246, 757—789 (1943).
Andrew, B. L.: A laryngeal pathway for aortic baroceptor impulses. J. Physiol. (Lond.) 125, 352—360 (1954a).
— Proprioception at the joint of the epiglottis of the rat. J. Physiol. (Lond.) 126, 507—523 (1954b).
— A functional analysis of the myelinated fibres of the superior laryngeal nerve. J. Physiol. (Lond.) 133, 420—432 (1956a).
— The nervous control of the cervical oesophagus of the rat during swallowing. J. Physiol. (Lond.) 134, 729—740 (1956b).
— Activity in afferent nerve fibres from the cervical oesophagus. J. Physiol. (Lond.) 135, 54—55 P (1957).
Aviado, D. M., and C. F. Schmidt: Reflex from stretch receptors in blood vessels, heart and lungs. Physiol. Rev. 35, 247—300 (1955).
Baumgarten, R. v., y L. Aranda Coddou: Distributión de las aferencias cardiovasculares y respiratorias en las raíces bulbares del nervio vago. Acta neurol. lat.-amer. 5, 267—278 (1959).

BAUMGARTEN, R. v., H. P. KOEPCHEN u. L. ARANDA: Untersuchungen zur Lokalisation der bulbären Kreislaufzentren. 1. Mitt. Funktionelle Organisation der Vagus-Wurzeln. Verh. Dtsch. Ges. Kreisl.-Forsch., 25. Tagg 1959, S. 170—172.

BEER, TH., u. A. KREIDL: Über den Ursprung der Vagusfasern, deren zentrale Reizung Verlangsamung, rsp. Stillstand der Atmung bewirkt. Pflügers Arch. ges. Physiol. 62, 156—165 (1896).

BEIN, H. J., u. H. HELMICH: Über afferente Vagusfasern. Helv. physiol. pharmacol. Acta 7, C40—C41 (1949).

—, u. R. MEIER: Pharmakologische Untersuchungen über Pendiomid, eine neuartige Substanz mit ganglienblockierender Wirkung. Schweiz. med. Wschr. 81, 446—452 (1951).

BIANCONI, R., and G. J. H. GREEN: Baroreceptor innervation of the bifurcation of the brachio-cephalic trunk in the cat. Arch. ital. Biol. 97, 47—52 (1959a).

— — Pulmonary baroreceptors in the cat. Arch. ital. Biol. 97, 305—315 (1959b).

— — Cardio-respiratory afferent fibres in the vagus of the cat. Arch. Sci. Biol. 43, 454—463 (1959c).

— — The right subclavian body. Arch. ital. Biol. 98, 1—9 (1960).

BLAIR, E. A., and J. ERLANGER: On the process of excitation by brief shocks in axons. Amer. J. Physiol. 114, 309—316 (1936).

BONHOEFFER, K., u. T. KOLATAT: Druckvolumendiagramm und Dehnungsreceptoren der Froschlunge. Pflügers Arch. ges. Physiol. 265, 477—484 (1958).

BONVALLET, M., et B. SIGG: Étude électrophysiologique des afférences vagales au niveau de leur pénetration dans le bulbe. J. Physiol. (Paris) 50, 63—74 (1958).

BOYD, J. D.: The nerve supply of the mammalian ductus arteriosus. J. Anat. (Lond.) 75, 457—468 (1941).

BRECHER, G. A.: Critical review of recent work on ventricular diastolic suction. Circulat. Res. 6, 554—566 (1958).

BRONK, D. W., and G. STELLA: Afferent impulses in the carotid sinus nerve. I. The relation of the discharge from single end organs to arterial blood pressure. J. cell. comp. Physiol. 1, 113—130 (1932).

— — The response to steady pressures of single end organs in the isolated carotid sinus. Amer. J. Physiol. 110, 708—714 (1935).

BROWN, C. L., and J. A. B. GRAY: Some effects of nicotine-like substances and their relation to sensory nerve endings. J. Physiol. (Lond.) 107, 306—317 (1948).

BRÜCKE, E.: Über ein in der Darmschleimhaut aufgefundenes Muskelsystem. S.-B. Akad. Wiss. Wien 6, 214—219 (1851).

BÜLBRING, E., and D. WHITTERIDGE: The activity of vagal stretch endings during congestion in perfused lungs. J. Physiol. (Lond.) 103, 477—487 (1945).

BYKOV, K. M.: Text-book of physiology. Moscow: Foreign Languages Publishing House 1958.

CADMAN, A. W.: The position of the respiratory and cardioinhibitory fibres in the rootlets of the IXth, Xth, and XIth cranial nerves. J. Physiol. (Lond.) 26, 42—47 (1900).

CAHOON, D. H., I. E. MICHAEL and V. JOHNSON: Respiratory modification of the cardiac output. Amer. J. Physiol. 133, 642—650 (1941).

CHAPMAN, K. M., and J. W. PEARCE: Vagal afferents in the monkey. Nature (Lond.) 184, 1237—1238 (1959).

CHASE, M. R., and S. W. RANSON: The structure of the roots, trunk and branches of the vagus nerve. J. comp. Neurol. 24, 31—60 (1914).

CHERNIGOVSKY, V. N.: Interoceptors. Moscow: Mediz 1960 [Russian].

CHRISTIE, R. V.: Dyspnoea: a review. Quart. J. Med. 31, 421—454 (1938).

CITTERS, R. L. VAN: Pulse wave velocity through polyethylene tubing. J. appl. Physiol. 15, 184—185 (1960).

COLERIDGE, J. C. G., A. HEMINGWAY, R. L. HOLMES and R. J. LINDEN: The location of atrial receptors in the dog: A physiological and histological study. J. Physiol. (Lond.) **136**, 174—197 (1957).

—, and C. KIDD: Electrophysiological evidence of baroreceptors in the pulmonary artery of the dog. J. Physiol. (Lond.) **150**, 319—331 (1960).

— — Relationship between pulmonary arterial pressure and impulse activity in pulmonary arterial baroreceptor fibres. J. Physiol. (Lond.) **158**, 197—205 (1961).

— — and J. A. SHARP: The distribution, connexions and histology of baroreceptors in the pulmonary artery, with some observations on the sensory innervation of the ductus arteriosus. J. Physiol. (Lond.) **156**, 591—602 (1961).

COSTANTIN, LEROY L.: Effect of pulmonary congestion on vagal afferent activity. Amer. J. Physiol. **196**, 49—53 (1959).

DALY, M. DEBURGH, and D. H. L. EVANS: Functional and histological changes in the vagus nerve of the cat after degenerative section at various levels. J. Physiol. (Lond.) **120**, 579—595 (1953).

DAVIS, H. L., W. S. FOWLER and E. H. LAMBERT: Effect of volume and rate of inflation and deflation on transpulmonary pressure and response of pulmonary stretch receptors. Amer. J. Physiol. **187**, 558—566 (1956).

DAWES, G. S.: Reflexes from the heart and lungs. Abstr. XIX. int. physiol. Congr. 1953, p. 51—59.

— Reflex factors in the regulation of the circulation. In: Shock and circulatory homeostasis. Transactions of the third Conference. New York: Josiah Macy Jr. Foundation 1954.

—, and J. H. COMROE jr.: Chemoreflexes from the heart and lungs. Physiol. Rev. **34**, 167—201 (1954).

—, and J. C. MOTT: Circulatory and respiratory reflexes caused by aromatic guanidines. Brit. J. Pharmacol. **5**, 65—76 (1950).

— — and J. G. WIDDICOMBE: Respiratory and cardiovascular reflexes from the heart and lungs. J. Physiol. (Lond.) **115**, 258—291 (1951).

DECASTRO, F.: Sur la structure de la synapse dans les chemorecepteurs: leur mécanisme d'excitation et rôle dans la circulation sanguine locale. Acta physiol. scand. **22**, 14—43 (1951).

DICKINSON, C. J.: Afferent nerves from the heart region. J. Physiol. (Lond.) **111**, 399—407 (1950).

DODT, E.: Differential thermosensitivity of mammalian A-fibres. Acta physiol. scand. **29**, 91—108 (1953).

DOUGLAS, W. W.: Is there chemical transmission at chemoreceptors? Pharmacol. Rev. **6**, 81—83 (1954).

—, and J. L. MALCOLM: The effect of localized cooling on conduction in cat nerves. J. Physiol. (Lond.) **130**, 53—71 (1955).

—, and J. M. RITCHIE: A technique for recording functional activity in specific groups of medullated and non-medullated fibres in whole nerve trunks. J. Physiol. (Lond.) **138**, 19—30 (1957a).

— — On excitation of non-medullated afferent fibres in the vagus and aortic nerves by pharmacological agents. J. Physiol. (Lond.) **138**, 31—43 (1957b).

— — and W. SCHAUMANN: Depressor reflexes from medullated and non-medullated fibres in the rabbit's aortic nerve. J. Physiol. (Lond.) **132**, 187—198 (1956).

DOWNING, S. E., and R. W. TORRANCE: Vagal baroreceptors of the bull-frog. J. Physiol. (Lond.) **156**, 13P (1961).

DUBOIS, F. S., and J. O. FOLEY: Experimental studies on the vagus and spinal accessory nerves in the cat. Anat. Rec. **64**, 285—307 (1936).

EAD, H. W., J. H. GREEN and E. NEIL: A comparison of the effects of pulsatile and non-pulsatile blood flow through the carotid sinus on the reflexogenic activity of the sinus baroceptors in the cat. J. Physiol. (Lond.) **118**, 509—519 (1952).

ELFTMANN, A. G.: The afferent and parasympathetic innervation of the lungs and trachea of the dog. Amer. J. Anat. **72**, 1—28 (1943).

ERLANGER, J.: Analysis of the action potential in nerve. From Harvey Lectures, 1936—1927, pp. 90—113. Baltimore: Williams & Wilkins Company 1928.

—, and H. S. GASSER: Electrical signs of nervous activity. Philadelphia: University of Pennsylvania Press 1937.

EVANS, D. H. L., and J. G. MURRAY: Histological and function studies on the fibre composition of the vagus nerve of the rabbit. J. Anat. (Lond.) **88**, 320—337 (1954a).

— — Regeneration of non-medullated nerve fibres. J. Anat. (Lond.) **88**, 465—480 (1954b).

EXNER, A.: Wie schützt sich der Verdauungstract vor Verletzungen durch spitze Fremdkörper. Pflügers Arch. ges. Physiol. **89**, 253—280 (1902).

EYZAGUIRRE, C., and K. UCHIZONO: Observations on the fibre content of nerves reaching the carotid body of the cat. J. Physiol. (Lond.) **159**, 268—281 (1961).

FOLEY, J. O., and F. S. Du BOIS: An experimental study of the rootlets of the vagus nerve in the cat. J. comp. Neurol. **60**, 137—159 (1934).

— — Quantitative studies of the vagus nerve in the cat. I. The ratio of sensory to motor fibres. J. comp. Neurol. **67**, 49—67 (1937).

GASSER, H. S.: The relation of the shape of the action potential of nerve to conduction velocity. Amer. J. Physiol. **84**, 699—711 (1928).

— Unmedullated fibres originating in dorsal root ganglia. J. gen. Physiol. **33**, 651—690 (1950).

—, and H. GRUNDFEST: Axon diameters in relation to the spike dimensions and the conduction velocity in mammalian A fibres. Amer. J. Physiol. **127**, 393—414 (1939).

GAUER, O. H., u. J. P. HENRY: Beitrag zur Homöostase des extraarteriellen Kreislaufs. Volumenregulation als unabhängiger physiologischer Parameter. Klin. Wschr. **34**, 356—366 (1956).

GERNANDT, B. Z., and Y. ZOTTERMAN: Intestinal pain: An electrophysiological investigation on mesenteric nerves. Acta physiol. scand. **12**, 56—72 (1946).

GRANIT, R.: Receptors and sensory perception. New Haven: Yale University Press 1955.

GRAY, J. A. B., and J. DIAMOND: Pharmacological properties of sensory receptors and their relation to those of the autonomic nervous system. Brit. med. Bull. **13**, 185—188 (1957).

—, and W. D. M. PATON: The circulation time in the cat, studied by a conductivity method. J. Physiol. (Lond.) **110**, 173—193 (1949).

GRUNDFEST, H., and H. S. GASSER: Properties of mammalian nerve fibers of slowest conduction. Amer. J. Physiol. **123**, 307—318 (1938).

GUNN, J. A., and S. W. F. UNDERHILL: Experiments on the surviving mammalian intestine. Quart. J. exp. Physiol. **8**, 275—296 (1914).

HAMBLETON, B. F.: Note upon the movements of the intestinal villi. Amer. J. Physiol. **34**, 446—447 (1914).

HAMMOUDA, M., and W. H. WILSON: The vagus influences giving rise to the phenomena accompanying expansion and collapse of the lungs. J. Physiol. (Lond.) **74**, 81—114 (1932).

HEAD, H.: On the regulation of respiration. J. Physiol. (Lond.) **10**, 1—70 (1889).

HEINBECKER, P., and J. O'LEARY: The mammalian vagus nerve; a functional and histological study. Amer. J. Physiol. **106**, 623—646 (1933).

HELLNER, K., u. R. v. BAUMGARTEN: Über ein Endigungsgebiet afferenter, kardiovasculärer Fasern des Nervus vagus im Rautenhirn der Katze. Pflügers Arch. ges. Physiol. **273**, 223—234 (1961).

HENRY, J. P., and J. W. PEARCE: The possible role of cardiac atrial stretch receptors in the induction of changes in urine flow. J. Physiol. (Lond.) **131**, 572—585 (1956).

Heymans, C., and E. Neil: Reflexogenic areas of the cardiovascular system. London: Churchill 1958.

Hoffman, H. H., and A. Kuntz: Vagus nerve components. Anat. Rec. 127, 551—567 (1957).

Holmes, R. L.: Structures in the atrial endocardium of the dog which stain with methylene blue, and the effects of unilateral vagotomy. J. Anat. (Lond.) 91, 259—266 (1957).

Hughes, R., A. J. May and J. G. Widdicombe: The effect of pulmonary congestion and oedema on lung compliance. J. Physiol. (Lond.) 142, 306—313 (1958).

Hunt, C. C.: The reflex activity of mammalian small-nerve fibres. J. Physiol. (Lond.) 115, 456—469 (1951).

— Relation of function to diameter in afferent fibers of muscle nerves. J. gen. Physiol. 38, 117—131 (1954).

—, and S. W. Kuffler: Stretch receptor discharges during muscle contraction. J. Physiol. (Lond.) 113, 298—315 (1951).

—, and A. K. McIntyre: An analysis of fibre diameter and receptor characteristics of myelinated cutaneous afferent fibres in cat. J. Physiol. (Lond.) 153, 99—112 (1960).

Hursh, J. B.: Conduction velocity and diameter of nerve fibers. Amer. J. Physiol. 127, 131—139 (1939).

Iggo, A.: Tension receptors in the stomach and the urinary bladder. J. Physiol. (Lond.) 128, 593—607 (1955).

— Gastro-intestinal tension receptors with unmyelinated afferent fibres in the vagus of the cat. Quart. J. exp. Physiol. 42, 130—143 (1957a).

— Gastric mucosal chemoreceptors with vagal afferent fibres in the cat. Quart. J. exp. Physiol. 42, 398—409 (1957b).

— The electrophysiological identification of single nerve fibres, with particular reference to the slowest-conducting vagal afferent fibres in the cat. J. Physiol. (Lond.) 142, 110—126 (1958).

Irisawa, H., A. P. Greer and R. F. Rushmer: Changes in the dimensions of the venae cavae. Amer. J. Physiol. 196, 741—744 (1959).

James, A. H.: The physiology of gastric digestion. London: Edward Arnold 1957.

Jarisch, A., and Y. Zotterman: Depressor reflexes from the heart. Acta physiol. scand. 16, 31—51 (1948).

Jones, R. L.: Components of the vagus nerve. Proc. Soc. exp. Biol. (N.Y.) 29, 1138—1141 (1932).

— Cell fibre ratios in the vagus nerve. J. comp. Neurol. 67, 469—482 (1937).

Katz, B.: Depolarization of sensory terminals and the initiation of impulses in muscle spindle. J. Physiol. (Lond.) 111, 261—282 (1950).

Keele, C. A., and E. Neil: Samson Wright's applied physiology. London: Oxford University Press 1961.

Keller, Ch. J., u. A. Loeser: Der zentripetale Lungenvagus. Z. Biol. 89, 373—395 (1926).

King, C. E., L. Arnold and J. G. Church: The physiological rôle of the intestinal mucosal movements. Amer. J. Physiol. 61, 80—92 (1922).

— L. C. Glass and S. E. Townsend: The circular components of the muscularis mucosae of the small intestine of the dog. Amer. J. Physiol. 148, 667—674 (1947).

—, and M. H. Robinson: The nervous mechanisms of the muscularis mucosae. Amer. J. Physiol. 143, 325—335 (1945).

Kiraly, J. K., and K. Krnjevic: Some retrograde changes in the function of nerves after peripheral section. Quart. J. exp. Physiol. 44, 244—257 (1959).

Knowlton, G. C., and M. G. Larrabee: Unitary analysis of pulmonary volume receptors. Amer. J. Physiol. 147, 100—114 (1946).

Kolatat, T., K. Kramer u. N. Mühl: Über die Aktivität sensibler Herznerven des Frosches und ihre Beziehungen zur Herzdynamik. Pflügers Arch. ges. Physiol. **264**, 127—144 (1957).

Kokas, E.: Vergleichend-physiologische Untersuchungen über die Bewegung der Darmzotten. Pflügers Arch. ges. Physiol. **225**, 416—420 (1930).

—, u. G. Ludány: Die Beobachtung der Zottenbewegung am überlebenden Darm. Pflügers Arch. ges. Physiol. **231**, 20—23 (1933).

Kramer, K.: Die afferente Innervation und die Reflexe von Herz und venösem System. Verh. Dtsch. Ges. Kreis.-Forsch. **25**. Tagg, 1959, S. 142—163.

Krayer, O.: The history of the Bezold-Jarisch effect. Naunyn-Schmiedeberg's Arch. exp. Path. Pharmak. **240**, 361—368 (1961).

— —, and G. H. Acheson: The pharmacology of the veratrum alkaloids. Physiol. Rev. **26**, 383—446 (1946).

Kreidl, A.: Die Wurzelfasern der motorischen Nerven des Oesophagus. Pflügers Arch. ges. Physiol. **59**, 9—16 (1895).

Krnjević, K., and N. M. van Gelder: Tension changes in crayfish stretch receptors. J. Physiol. (Lond.) **159**, 310—325 (1961).

Kuffler, S. W., C. C. Hunt and J. P. Quilliam: Function of medullated small-nerve fibers in mammalian ventral roots: efferent spindle innervation. J. Neurophysiol. **14**, 29—54 (1951).

Landgren, S.: On the excitation mechanism of the carotid baroceptors. Acta physiol. scand. **26**, 1—34 (1952a).

— The baroceptor activity in the carotid sinus nerve and the distensibility of the sinus wall. Acta physiol. scand. **26**, 35—56 (1952b).

Langrehr, D.: Entladungsmuster und allgemeine Reizbedingungen von Vorhofsreceptoren bei Hund und Katze. Pflügers Arch. ges. Physiol. **271**, 257—269 (1960a).

— Beziehungen zwischen Vorhofsreceptoraktivitäten und Herzmechanik von Hund und Katze bei verschiedenen Kreislaufzuständen. Pflügers Arch. ges. Physiol. **271**, 270—282 (1960b).

—, u. K. Kramer: Beziehungen der mittleren Impulsfrequenz von Vorhofsreceptoren zum thorakalen Blutvolumen. Pflügers Arch. ges. Physiol. **271**, 797—807 (1960).

Larrabee, M. G., and G. C. Knowlton: Excitation and inhibition of phrenic motoneurones by inflation of the lungs. Amer. J. Physiol. **147**, 90—99 (1946).

Larsell, O.: Nerve terminations in the lung of the rabbit. J. comp. Neurol. **33**, 105—131 (1921).

Lemere, F.: Innervation of the larynx I. Innervation of laryngeal muscles. Amer. J. Anat. **51**, 417—437 (1932).

Little, R. C.: Volume elastic properties of the right and left atrium. Amer. J. Physiol. **158**, 237—240 (1949).

— Volume pressure relationships of the pulmonary-left heart vascular segment. Circulat. Res. **8**, 594—599 (1960).

Lloyd, D. P. C.: Neuron patterns controlling transmission of ipsilateral hind limb reflexes in cat. J. Neurophysiol. **6**, 293—315 (1943).

Lundberg, A.: Potassium and the differential thermosensitivity of membrane potential, spike, and negative after potential in mammalian A & C fibres. Acta physiol. scand. **15**, Suppl. 50, 1—67 (1948).

Marshall, R., and J. G. Widdicombe: The activity of pulmonary stretch receptors during congestion of the lungs. Quart. J. exp. Physiol. **43**, 320—330 (1958).

Matthews, B. H. C.: Nerve endings in mammalian muscle. J. Physiol. (Lond.) **78**, 1—53 (1933).

Meier, R., u. J. J. Bein: Neuere Befunde über die organisationsspezifischen Wirkungen am autonomen Nervensystem. Bull. schweiz. Akad. med. Wiss. **6**, 209—233 (1950).

Mott, J. C., and A. S. Paintal: The action of 5-hydroxytryptamine on pulmonary and cardiovascular vagal afferent fibres and its reflex respiratory effects. Brit. J. Pharmacol. 8, 238—241 (1953).

Mühl, N., I. Scholderer u. K. Kramer: Über die Aktivität der intrathorakalen Gefäß-rezeptoren und ihre Beziehung zur Herzfrequenz bei Änderung des Blutvolumens. Verh. Dtsch. Ges. Kreis.-Forsch. 22. Tagg, 1956, S. 122—126.

Murray, J. G.: Innervation of the intrinsic muscles of the cat's larynx by the recurrent laryngeal nerve: a unimodal nerve. J. Physiol. (Lond.) 135, 206—212 (1957).

Neil, E., and N. Joels: The impulse activity in cardiac afferent vagal fibres. Naunyn-Schmiedeberg's Arch. exp. Path. Pharmak. 240, 453—460 (1961).

— L. Ström and Y. Zotterman: Action potential studies of afferent fibres in the IXth and Xth cranial nerves of the frog. Acta physiol. scand. 20, 338—350 (1950).

—, and Y. Zotterman: Cardiac vagal afferent fibres in the cat and the frog. Acta physiol. scand. 20, 160—165 (1950).

Nonidez, J. F.: Identification of the receptor areas in the venae cavae and pulmonary veins which initiate reflex cardiac acceleration (Bainbridge's reflex). Amer. J. Anat. 61, 203—231 (1937).

— Studies on the innervation of the heart. II. Afferent nerve endings in the large arteries and veins. Amer. J. Anat. 68, 151—189 (1941).

Ogura, J. H., and R. L. Lam: Anatomical and physiological correlations on stimulating the human superior laryngeal nerve. Laryngoscope (St. Louis) 63, 947—959 (1953).

Opdyke, D. F., J. Duomarco, W. H. Dillon, H. Schreiber, R. C. Little and R. D. Seely: Study of simultaneous right and left atrial pressure pulses under normal and experimentally altered conditions. Amer. J. Physiol. 154, 258—272 (1948).

— H. F. van Noate and G. A. Brecher: Further evidence that inspiration increases right atrial inflow. Amer. J. Physiol. 162, 259—265 (1950).

Paintal, A. S.: The study of respiratory and cardiovascular reflex mechanisms involving the lungs. Ph. D. Thesis, Edinburgh 1952.

— A study of right and left atrial receptors. J. Physiol. (Lond.) 120, 596—610 (1953a).

— The response of pulmonary and cardiovascular receptors to certain drugs. J. Physiol. (Lond.) 121, 182—190 (1953b).

— The conduction velocities of respiratory and cardiovascular afferent fibres in the vagus nerve. J. Physiol. (Lond.) 121, 341—359 (1953c).

— Impulses in vagal afferent fibres from stretch receptors in the stomach and their role in the peripheral mechanism of hunger. Nature (Lond.) 172, 1194—1195 (1953d).

— A method of locating the receptors of visceral afferent fibres. J. Physiol. (Lond.) 124, 166—172 (1954a).

— A study of gastric stretch receptors. Their role in the peripheral mechanism of satiation of hunger and thirst. J. Physiol. (Lond.) 126, 255—270 (1954b).

— The response of gastric stretch receptors and certain other abdominal and thoracic vagal receptors to some drugs. J. Physiol. (Lond.) 126, 271—285 (1954c).

— Impulses in vagal afferent fibres from specific pulmonary deflation receptors. The response of these receptors to phenyl diguanide, potato starch, 5-hydroxytryptamine and nicotine, and their role in respiratory and cardiovascular reflexes. Quart. J. exp. Physiol. 40, 89—111 (1955a).

— The study of ventricular pressure receptors and their role in the Bezold reflex. Quart. J. exp. Physiol. 40, 348—363 (1955b).

— A method of recording the pulmonary circulation times in the cat. From: Proc. Inter. Conf. Peaceful Uses of Atomic Energy, vol. 12, p. 278—280. New York: United Nations 1956a.

— Excitation of sensory receptors in the thoracic and abdominal viscera. Abstr. XX. int. physiol. Congr. 1956b, p. 78—89.

— The location and excitation of pulmonary deflation receptors by chemical substances. Quart. J. exp. Physiol. 42, 56—71 (1957a).

PAINTAL, A. S.; The influence of certain chemical substances on the initiation of sensory discharges in pulmonary and gastric stretch receptors and atrial receptors. J. Physiol. (Lond.) **135**, 486—510 (1957b).
— Responses from mucosal mechanoreceptors in the small intestine of the cat. J. Physiol. (Lond.) **139**, 353—368 (1957c).
— Intramuscular propagation of sensory impulses. J. Physiol. (Lond.) **148**, 240—251 (1959).
— Functional analysis of group III afferent fibres of mammalian muscles. J. Physiol. (Lond.) **152**, 250—270 (1960).
— Determination of intrathoracic conduction time in cardiovascular afferent fibres of the vagus nerve. J. Physiol. (Lond.) **163**, 222—238 (1962).
PARTRIDGE, R. C.: Afferent impulses in the vagus nerve. J. cell. comp. Physiol. **2**, 367—380 (1933).
PEARCE, J. W.: The control of pulmonary blood pressure and its relation to certain reflex respiratory phenomena. D. Phil. Thesis, Oxford 1951.
— J. P. HENRY and K. M. CHAPMAN: The behaviour and possible functions of cardiac atrial stretch receptors. Abstr. XX. int. physiol. Congr. 1956, p. 711—712.
—, and D. WHITTERIDGE: The relation of pulmonary arterial pressure variations to the activity of afferent pulmonary vascular fibres. Quart. J. exp. Physiol. **36**, 177—188 (1951).
PÓRSZÁSZ, J., GY. SUCH, I. MARARASZ, M. BERTA and K. GIBISZER PÓRSZÁSZ: Possibility of summation in the vasomotor and respiratory centre, electrophysiological investigations on the vagal nerve in the cat. Acta physiol. Acad. Sci. hung. **17**, 23—34 (1960).
RANSON, S. W., J. O. FOLEY and C. D. ALPERT: Observations on the structure of the vagus nerve. Amer. J. Anat. **53**, 289—315 (1933).
—, and P. MIHALIK: The structure of the vagus nerve. Anat. Rec. **54**, 355—360 (1932).
SANDERS, F. K., and D. WHITTERIDGE: Conduction velocity and myelin thickness in regenerating nerve fibres. J. Physiol. (Lond.) **105**, 152—174 (1946).
SCHAEFER, H.: Elektrophysiologie der Herznerven. Ergebn. Physiol. **46**, 71—125 (1950).
SCHNEIDER, J. A., and F. F. YONKMAN: Action of serotonin (5-hydroxytryptamine) on vagal afferent impulses in the cat. Amer. J. Physiol. **174**, 127—134 (1953).
— — Species differences in the respiratory and cardiovascular response to serotonin (5-hydroxytryptamine). J. Pharmacol. exp. Ther. **111**, 84—98 (1954).
STÄMPFLI, R.: Bau und Funktion isolierter markhaltiger Nervenfasern. Ergebn. Physiol. **47**, 70—165 (1952).
STRUPPLER, A.: Afferente vagale Herznervenimpulse und ihre Beziehung zur Hämodynamik. Z. Biol. **107**, 416—428 (1955).
—, u. E. STRUPPLER: Über spezielle Charakteristika afferenter vagaler Herznervenimpulse und ihre Beziehungen zur Herzdynamik. Acta physiol. scand. **33**, 219—231 (1955).
SUCH, GY., and J. PÓRSZÁSZ: On the reversal of the vagal vasomotor reflexes. Acta physiol. Acad. Sci. hung. **17**, 35—38 (1960).
SWAN, A. A. B., and D. WHITTERIDGE: Baroreceptor fibres from the pulmonary artery. Abstr. XX. int. physiol. Congr. 1956, p. 867—868.
TAKINO, M., u. S. WATANABE: Über die Bedeutung des Ligamentum arteriosum bzw. des Ductus Botalli und der Ansatzstelle desselben an der Pulmonalwand (A. pulm.) als Blutdruckzügler bei verschiedenen Tierarten. Arch. Kreisl.-Forsch. **11**, 18—27 (1937).
TASAKI, I.: Nervous transmission. Springfield: Ch. C. Thomas 1953.
TORRANCE, R. W., and D. WHITTERIDGE: Technical aids in the study of respiratory reflexes. J. Physiol. (Lond.) **107**, 6—7 P (1948).

Tower, S. S.: Action potentials in sympathetic nerves, elicited by stimulation of frog's viscera. J. Physiol. (Lond.) **78**, 225—245 (1933).

Verzár, F., u. E. Kokas: Die Rolle der Darmzotten bei der Resorption. Pflügers Arch. ges. Physiol. **217**, 397—412 (1927).

Walsh, E. G., and D. Whitteridge: Vagal activity and the tachypnoea produced by multiple pulmonary emboli. J. Physiol. (Lond.) **103**, 37—38 P (1945).

Weidmann, H., B. Berde u. K. Bucher: Die Lage der vagalen Dehnungsrezeptoren in der Lunge. Helv. physiol. pharmacol. Acta **7**, 476—481 (1949).

—, u. K. Bucher: Zur Frage der Spezifität der vagalen Dehnungsreceptoren in der Lunge. Helv. physiol. pharmacol. Acta **9**, 94—100 (1951).

Wells, H. S., and R. G. Johnson: The intestinal villi and their circulation in relation to absorption and secretion of fluid. Amer. J. Physiol. **109**, 387—402 (1934).

Whitteridge, D.: Afferent impulses from the heart and lungs. Abstr. XVII. int. physiol. Congr. 1947.

— Afferent nerve fibres from the heart and lungs in the cervical vagus. J. Physiol. (Lond.) **107**, 496—512 (1948).

— Multiple embolism of the lung and rapid shallow breathing. Physiol. Rev. **30**, 475—486 (1950).

— Electrophysiology of afferent cardiac and pulmonary fibres. Abstr. XIX. int. physiol. Congr. 1953, p. 66—72.

— Effects of anaesthetics on mechanical receptors. Brit. med. Bull. **14**, 5—7 (1958).

—, and E. Bülbring: Changes in activity of pulmonary receptors in anaesthesia and their influence on respiratory behaviour. J. Pharmacol. exp. Ther. **81**, 340—359 (1944).

Widdicombe, J. G.: Respiratory reflexes from the trachea and bronchi of the cat. J. Physiol. (Lond.) **123**, 55—70 (1954a).

— Receptors in the trachea and bronchi of the cat. J. Physiol. (Lond.) **123**, 71—104 (1954b).

— Respiratory reflexes excited by inflation of the lungs. J. Physiol. (Lond.) **123**, 105—115 (1954c).

— The site of pulmonary stretch receptors in the cat. J. Physiol. (Lond.) **125**, 336—351 (1954d).

— Action potentials in vagal afferent nerve fibres to the lungs of the cat. Naunyn-Schmiedeberg's Arch. exp. Path. Pharmak. **241**, 415—432 (1961).

Wiggers, C. J.: The pressure pulses in the cardiovascular system. London: Longmans Green & Co. 1928.

Witzleb, E.: Zur Frage von cholinergischen Mechanismen bei der Erregung von afferenten Systemen. Pflügers Arch. ges. Physiol. **269**, 439—470 (1959).

Wyss, O. A. M.: The part played by the lungs in the reflex control of breathing. Helv. physiol. pharmacol. Acta **12**, Suppl. 10, 26—35 (1954).

—, et A. Rivkine: Les fibres afférentes du nerf vague participant aux réflexes respiratoires. Helv. physiol. pharmacol. Acta **8**, 87—106 (1950).

Zotterman, Y.: Touch, pain and tickling: an electro-physiological investigation on cutaneous sensory nerves. J. Physiol. (Lond.) **95**, 1—28 (1939).

Psychophysiologische Leistungsfähigkeit des Macacus-Affen nach Cortexausschaltungen

Von

K. BÄTTIG* und H. E. ROSVOLD**

Mit 5 Abbildungen

Inhaltsverzeichnis

I. Einleitung

Die alte Vorstellung, daß der Assoziationscortex, auch höherer oder sekundärer Cortex genannt, das spezifische Organ der höchsten nervösen Leistungen des Lernens, Assoziierens und Denkens darstelle, darf heute als überholt angesehen werden. Ebensogut wie solche Funktionen auch an die Intaktheit anderer Hirnstrukturen gebunden sind, hat der sekundäre Cortex auch Funktionen, die mit höheren psychischen Funktionen nichts zu tun haben.

Doch verbleibt dieser Teilaspekt der Psychofunktion weiterhin im Interesse und Blickpunkt der modernen Cortexforschung. Im letzten Jahrhundert führten Beobachtungen von Naturwissenschaftlern und Ärzten zu verschiedenen Deutungsversuchen der psychofunktionellen Organisation des Cortex, die den späteren Erkenntnissen nicht standhalten konnten. Wenn die neueren Arbeiten der 30—50 letzten Jahre in diese Fragen mehr Klarheit bringen konnten, ist dies weitgehend dem Umstand zu verdanken, daß die heute

---

* Institut für Hygiene und Arbeitsphysiologie der Eidg. Techn. Hochschule Zürich (Direktor: Prof. Dr. med. E. GRANDJEAN).

** Chief, Section on Animal Behavior, Laboratory of Psychology, National Institute of Mental Health, National Institutes of Health, Bethesda, U.S.A.

verfeinerten und ausgebauten Methoden der experimentellen Psychologie an Stelle der reinen Beobachtung Eingang in die Physiologie genommen haben.

Drei Hauptklassen psychischer Leistungen sind Gegenstand der modernen Psychophysiologie:

1. Nicht erlernte Reaktionen auf Umweltreize, zu denen das Instinktverhalten allgemein, Spontanaktivität, tropistisches Verhalten usw. gehören.

2. Erlernte Reaktionen auf Umweltreize. Hierzu gehören alle Reaktionen, die sich als „bedingte Reflexe" erklären lassen, also Reaktionen, die ein Tier als Anpassung an eine spezielle Umweltbedingung erworben hat, und die wieder erlöschen, sobald diese speziellen Umweltbedingungen nicht mehr bestehen.

3. „Symbolische Reaktionen". Darunter sind Reaktionen zu verstehen, die nicht als Antwort auf einen Umweltreiz ausgeführt werden, sondern die einem endogenen neuralen Reiz des Tieres entspringen. Der endogene Reiz substituiert also den nicht vorhandenen äußeren Reiz, er ist ein „symbolischer" Reiz. Hierher gehören z.B. beim Menschen Gedanken, Vorstellung, Sprache usw. Die Bezeichnungen Intelligenz, Einsicht, Erwartung usw. stellen Synonyme für diese Klasse psychischen Geschehens dar.

Bei der Suche nach Beziehungen zwischen diesen drei Klassen psychischer Funktionen und dem Aufbau und der physiologischen Funktionsweise des nervösen Apparates sind prinzipiell drei methodische Wege denkbar:

1. Die experimentelle Auslösung von Reaktionen durch adäquate Reizung von neuralen Strukturen. Dieser Weg hat sich bisher als sehr fruchtbar erwiesen in der Erforschung der 1. Klasse psychischer Tätigkeiten. Mit dieser Methode Beziehungen zu erlerntem oder gar symbolischem Verhalten herzustellen stieß jedoch bis heute auf große Schwierigkeiten.

2. Das Studium psychischer Vorgänge als Funktion physiologischer Vorgänge im Zentralnervensystem. Dieser Weg wäre der direkteste, jedoch sind die Kenntnisse des normalen physiologischen Geschehens im Nervensystem noch zu wenig ausgebaut. Es wurden zwar Beziehungen zwischen psychischen Prozessen der 1. Klasse (besonders Weckreaktion, Schreckreaktion, Alarmreaktion usw.) mit elektroencephalographischen Vorgängen nachgewiesen, und neuerdings konnten auch psychische Prozesse der zweiten Ordnung (bedingte Reaktionen) mit charakteristischen Hirnstrombildern in Parallele gesetzt werden. Dagegen fehlen Beziehungen von elektrischen Vorgängen im Hirn zu psychischen Vorgängen der höchsten dritten Ordnung. Ferner steht das weite Feld eventueller Beziehungen zwischen biochemischen Reaktionsabläufen und psychischen Prozessen im allerersten Beginn der systematischen Erforschung.

3. Das Studium von Funktionsausfällen nach selektiver Entfernung von Bestandteilen des nervösen Apparates. Diese Methode erbrachte Beziehungen zwischen dem Zentralnervensystem und allen drei Klassen psychischer Ver-

haltens. Die Ausschaltungstechnik hat aber verschiedene Nachteile. Sie beschädigt nicht nur die eine sorgfältig untersuchte Funktion, sondern auch andere und nicht psychische Funktionen. Sie stellt eine Amputation des Nervensystems dar, dessen Folgen der Organismus ganz oder teilweise kompensiert, so daß das Ergebnis einer Untersuchung nach der Operation in eine Resultante zwischen dieser Kompensation und der eigentlichen Direktfolge des Eingriffes ausmündet. Ferner geht die Ausschaltungstechnik oft von einer zu starren anatomischen Konzeption der funktionellen Struktur des nervösen Apparates aus, und die Gefahr, daß bei diesem Vorgehen dynamische Aspekte der Hirnfunktion übersehen werden, ist groß.

Andererseits können diese Ausschaltungsstudien im Tierversuch unser Interesse direkt beanspruchen, weil sie sich unmittelbar zu den Gegebenheiten der Klinik in Parallele setzen lassen, wo chirurgischer oder pathologischer Ausfall von Hirngewebe den gleichen Fragenkomplex stellen. In dieser Übersicht sollen Ausschaltungsstudien am Macacus-Affen besprochen werden, wobei vorwiegend auf die Wirkung von Ausschaltungen auf die Assoziationsfähigkeit und höhere psychische Leistungen eingegangen werden soll.

Allgemeine Methodik

Versuchsplanung. Den meisten Studien lag folgender Plan zugrunde: Wenn eine Ausschaltung I die Funktion A beschädigt, nicht aber die Funktion B, während eine Ausschaltung II das Umgekehrte tut, darf auf die Verschiedenwertigkeit der beiden getesteten Cortexfelder geschlossen werden. Daher umfaßten die meisten der Studien mehrere Gruppen zu durchschnittlich etwa vier Tieren. Jede Gruppe erhielt eine andere Ausschaltung, und alle Tiere wurden mit den gleichen psychologischen Testen verglichen.

Chirurgische und anatomische Methoden. Allgemein wurden die Tiere in Narkose und unter aseptischen Bedingungen operiert. Nach Beendigung aller psychologischen Teste wurden die Tiere getötet, die Hirne entfernt und die chirurgisch gesetzten Ausschaltungen histologisch kontrolliert. Als Hinweis auf das Ausmaß der gesetzten Ausschaltungen dient für die nachfolgende Besprechung der einzelnen Studien die in der Abb. 1 wiedergegebene cytoarchitektonische Rindenkarte des Macaca mulatta-Affen nach v. BONIN und BAILEY.

In diese Karten wurden für die einzelnen Felder auch die klassischen Nummernbezeichnungen von BRODMANN eingetragen.

Bei einem Teil der in dieser Arbeit besprochenen Ausschaltungen waren von den Autoren annähernd dieselben Begrenzungen und derselbe Umfang geplant worden, so daß man von eigentlichen „Standardausschaltungen" sprechen kann. Die vier wichtigsten dieser Ausschaltungen sind in der Abb. 2 schematisch dargestellt. Es handelt sich dabei um die „Frontalausschaltung" (F), die

,,Parietalausschaltung" (*P*), die ,,Occipitalausschaltung" (*O*) und die ,,Temporalausschaltung" (*T*). Soweit im folgenden diese Bezeichnungen verwendet werden, wird dabei anatomisch an Felder gedacht, wie sie auf dieser Abb. 2 eingetragen sind. Die Begrenzungen dieser Felder lassen sich wie folgt beschreiben: Die ,,Frontalausschaltung" beschränkt sich auf die laterale Fläche des Frontal-

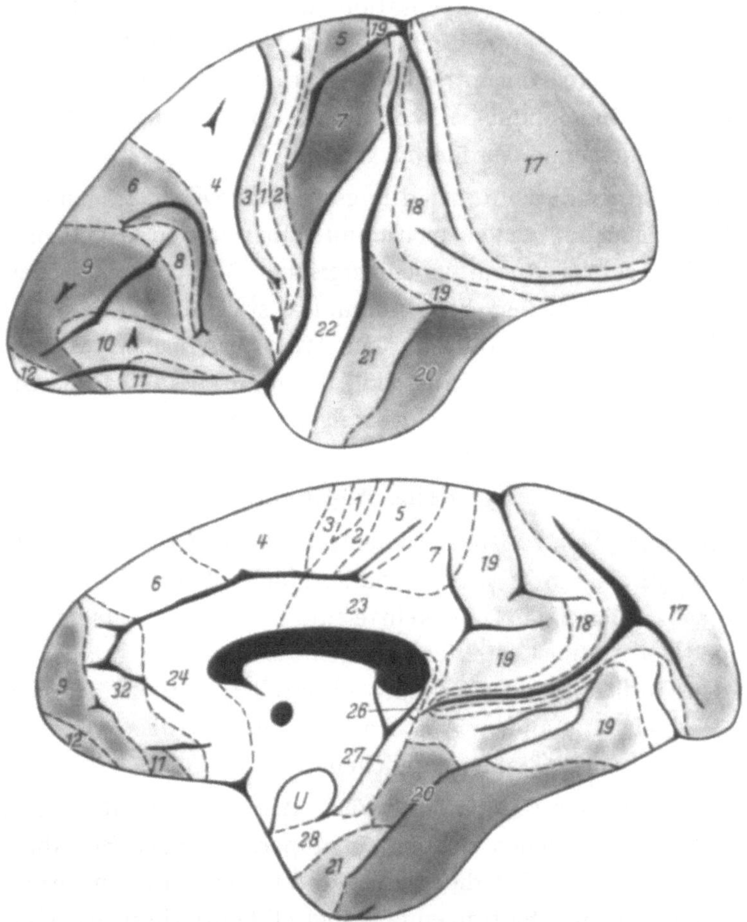

Abb. 1.   Cytoarchitektonische Darstellung des Cortex des Macaca mulatta-Affen nach von Bonin und Bailey (4); oben: laterale Ansicht; unten: mediale Ansicht

lappens inklusive den Pol, ohne jedoch auf die ventrale oder mediale Fläche des Lappens überzugreifen. Caudalwärts geht sie bis zur Grenzlinie zwischen den Brodmann-Feldern 4 und 6 gemäß Abb. 1. Die ,,Parietalausschaltung" umfaßt den lateralen Anteil des Parietallappens und ist rostral durch den Sulcus intraparietalis und caudal durch Sulcus lunatus und Sulcus temporalis superior begrenzt. Die ,,Occipitalausschaltung" besteht im ganzen lateralen Anteil des Feldes Nr. 17 gemäß der Abb. 1. Die ,,Temporalausschaltung" endet etwa 2 cm vor dem Pol des Temporallappens. Sie geht caudalwärts bis zur Vena v. Labbé, dorsalwärts bis zum Sulcus temporalis superior und erfaßt auf der ventralen Fläche des Temporallappens die dort liegenden Anteile des Feldes Nr. 20 nach BRODMANN gemäß der Abb. 1.

Psychologischer Testapparat. Die meisten Autoren verwenden für ihre Untersuchungen den unter dem Namen „Wisconsin" General Testing Apparatus" (WGTA) bekannten einfachen Apparat in verschiedensten Modifikationen.

Der Apparat (Abb. 3) besteht im wesentlichen aus einer Wahlkammer mit Futterbehältern. Vorder- und Hinterwand dieser Wahlkammer sind als auf- und abbewegliche Schiebetore ausgebildet. Der getestete Affe sitzt auf der einen Seite der Kammer in seinem Käfig, der Experimentator auf der anderen Seite, wo er mittels Seilzügen die beiden Schiebetore wahlweise so stellen kann, daß entweder er selber, oder der Affe Zugang zur Wahlkammer hat. Er gibt sich jeweils selber Zugang zur Kammer, um eine Testsituation vorzubereiten, indem er z. B. in

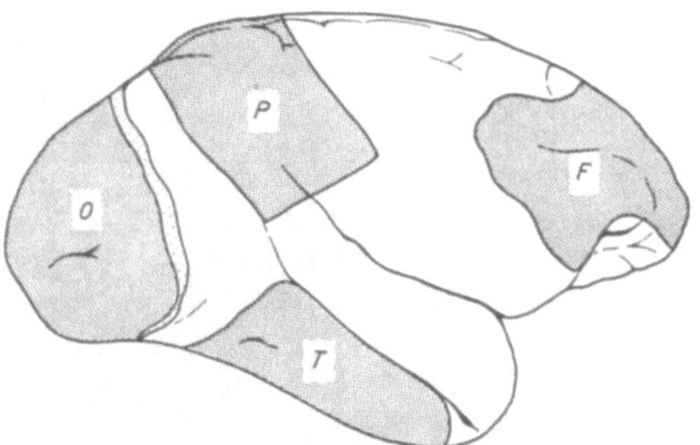

Abb. 2. Schematische Darstellung von 4 „Standardläsionen"

einem bestimmten Futterbehälter Futter versteckt. Das Tier kann diesen Vorgang nicht mitansehen, weil der Schieber von seiner Seite aus undurchsichtig ist. Hat aber das Tier Zugang zu der Wahlkammer, so kann der Experimentator die Reaktion des Tieres mitansehen, weil sein Schieber semitransparent ist.

In der Zukunft dürfte die Verwendung vollautomatischer Testapparate der funktionellen Cortexforschung beim Tier neue Möglichkeiten erschließen. Das Prinzip der Automatisierung wird bei der allgemein bekannten Skinner-Technik des „operant condition" bereits seit Jahren angewendet und hat sich vor allem in der experimentellen Verhaltensforschung und in der Psychopharmakologie als fruchtbar erwiesen.

Testung von nicht erlerntem Verhalten. Der WGTA wird verwendet, um die Sehschärfe und Haltung des Tieres gegenüber unbekannten neuen Objekten, die im Apparat ausgelegt werden, zu testen. Dagegen werden soziales, emotionelles und sexuelles Verhalten meist in der freien Gruppen- oder Einzelsituation beobachtet und nach einem willkürlich gewählten Punktesystem bewertet. Spontane Aktivität wird meist in einem normalen Aufenthaltskäfig gemessen, indem vermittels Selenzellen gezählt wird, wie oft in einem bestimmten Zeitraum ein Tier einen Lichtstrahl kreuzt.

Testung des assoziativen Lernen. Im Prinzip wird dem Tier im WGTA eine Situation so oft präsentiert, bis es lernt, daß bei einem bestimmten Reiz Futter nur im einen und nicht im anderen Behälter zu finden ist.

Bei der einfachen „Simultanwahl" enthält die Wahlkammer des WGTA mindestens zwei Futterbehälter oder Futterlöcher. Auf eines der beiden, bald links, bald rechts, wird das eine Merkmal gelegt, unter welchem immer Futter versteckt ist, während unter dem anderen Merkmal kein Futter verborgen ist.

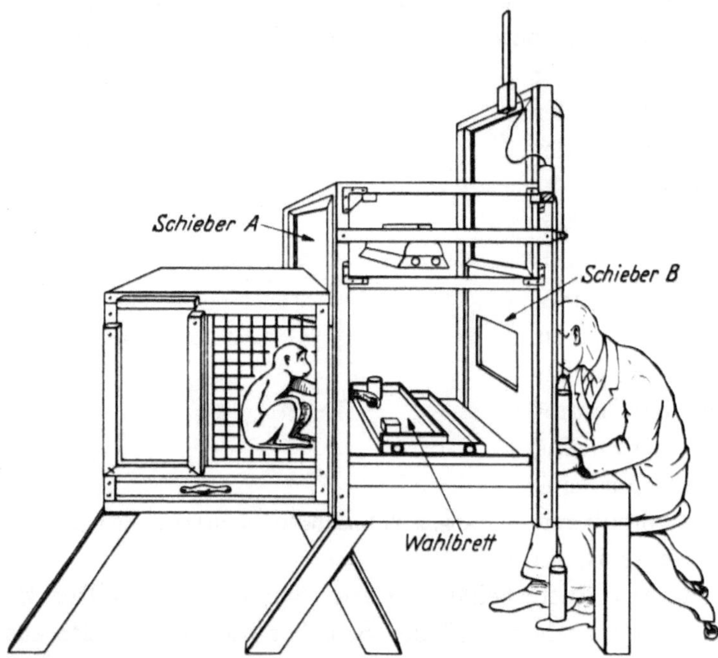

Abb. 3. „Wisconsin General Testing Apparatus" (WGTA), nach H. F. Harlow und P. H. Settlage (33). Der Schieber A ist undurchsichtig. Der Schieber B ist semitransparent und erlaubt dem Beobachter die Kontrolle des Wahlbrettes, ohne vom Tier gesehen zu werden

Nach einer Anzahl von Versuchstagen zu je etwa 20—50 Einzelversuchen wird ein normales Tier je nach dem Schwierigkeitsgrad der Unterscheidung lernen, daß es das Futter immer nur unter einem der beiden Merkmale suchen muß. Auf diese Art kann die Unterscheidung von Gesichts-, Geruchs- und Tastmerkmalen geprüft werden.

Gehörsmerkmale können dagegen nicht dem einen oder dem anderen Futterbehälter zugeordnet werden. Die Lösung besteht hier darin, dem Tier nur einen Unterscheidungston pro Einzelversuch zu geben, und für diesen Ton das Futter immer in den einen Futterbehälter zu legen, und für den anderen Unterscheidungston immer in den anderen Behälter. Bei dieser Testart sind beide Futterbehälter also neutral, nicht mit einem Unterscheidungsmerkmal versehen. Das Tier muß je nach dem Reiz den einen oder anderen Behälter wählen. Diese besondere Testvariante wird simultane „Bedingungswahl" genannt (Wahl je nach der Bedingung). Visuelle, Geruchs- oder Tastreize können natürlich ebenfalls nach dieser Methode antrainiert werden.

Bei der „Sukzessivwahl" ist auf dem Wahlbrett nur ein Futterbehälter vorhanden und es wird pro Versuch nur der Reiz A oder B gegeben. Der Reiz A bedeutet immer, daß Futter in dem einzigen vorhandenen Futterloch ausgelegt ist. Der Reiz B bedeutet, daß kein Futter in dem Loch vorhanden ist. Muß das Tier bei der „Simultanwahl" lernen, von mehr als einem Futterbehälter den korrekten zu öffnen, so muß es bei der Sukzessivwahl lernen, je nach Reiz Futter zu erwarten oder nicht, also den einzigen vorhandenen Futterbehälter bald zu öffnen, bald nicht zu öffnen.

Im WGTA sind die beiden Testarten in ihrer Schwierigkeit anscheinend verschieden. Die Simultanwahl wird im allgemeinen rascher erlernt. Auch beobachtet man bei der Simultanwahl, daß ein Tier nach der ersten Zeit des erfolglosen Ratens „wo ist das Futter?" plötzlich korrekt reagiert, also im Sinne eines „Aha-Erlebnisses". Bei der Sukzessivwahl dagegen setzt sich der Erfolg nur nach und nach durch, das Tier lernt dabei graduell, auf den unbelohnten Reiz nicht mehr zu reagieren.

Testung symbolischer Reaktionen. Für diese Gruppe von psychischen Leistungen wird wiederum zum größten Teil der WGTA verwendet.

Aufgeschobene Reaktion: Im Prinzip wird zwischen dem Reiz zur Reaktion und der Reaktion selber ein zeitlicher Aufschub eingeschoben. Praktisch geschieht dies so, daß beide Schirme des WGTA gehoben werden, während das Futter in eines der beiden Futterlöcher gelegt wird. Der Affe kann also zusehen, an welchem Ort das Futter versteckt wird. Dann wird der Schirm vor dem Tier gesenkt und beide Futterlöcher werden mit neutralen Kartons zugedeckt. Erst Sekunden bis Minuten später hebt der Experimentator den Schirm vor dem Tier wieder. Das Tier muß also die Zeit des Aufschubes durch einen neuralen „symbolischen Prozeß" überbrücken, denn der physische Reiz ist im Moment der Reaktion ja nicht mehr gegenwärtig. Dauer des Aufschubes und Testart dieser Aufgabe können in verschiedener Weise variiert werden. So kann die Aufgabe „direkt" gegeben werden, indem dem Tier vor dem Aufschub die Belohnung selber gezeigt wird, oder „indirekt" wenn statt dem Futter dem Tier über dem korrekten Futterloch ein entsprechendes Symbol für das Futter gezeigt wird. Ferner kann der Test nach dem Prinzip der „Simultanwahl" oder nach dem Prinzip der „Sukzessivwahl" ausgeführt werden. Im ersten Fall wählt das Tier nach dem Aufschub den linken oder rechten Futterplatz. Im zweiten Fall greift es je nach dem vorausgegangenen Reiz nach dem Futterversteck, oder unterläßt eine Reaktion überhaupt. Mit solchen Varianten läßt sich für das Tier der Grad der Schwierigkeit variieren.

Aufgeschobene Alternation: Bei diesem Aufschubtest wird dem Tier vor dem Aufschub kein Reiz geboten. Statt dessen liegt die Belohnung rein abwechslungsweise einmal im linken und einmal im rechten Futterloch. Der

Affe muß also lernen, für jede Reaktion das Ergebnis des vorangegangenen Versuches als Reiz zu benützen. Auch diese Aufgabe kann in den ähnlichen Varianten wie die aufgeschobene Reaktion gestellt werden.

Testung von „Hypothesen" des Tieres: Dem Tier wird ein schwieriges oder unlösbares Problem gestellt. Jede Reaktion wird notiert und gleiche Reaktionen werden zu Gruppen zusammengestellt. Macht das Tier vorerst eine Serie von Reaktionen des Typs A, repräsentieren diese Versuche die erste „Hypothese" oder „Erwartung" des Tieres; eine zweite darauffolgende Gruppe von Reaktionen des Typs B repräsentiert die „zweite Hypothese" des Tieres usw. Hat das Tier keine „Hypothesen", so sind seine Lösungsversuche unsystematisch, zufällig und nicht durch einen aktiven zentralen Prozeß des „Überlegens" bestimmt.

Serielernen: Bei diesen Testen erhält das Tier hintereinander eine ganze Reihe von Unterscheidungsaufgaben gestellt (visuelle oder andere Probleme). Normalerweise wird jedes der Einzelprobleme dem Tier nur in einer bestimmten Zahl von Einzelversuchen gestellt, z.B. jedes Problem in nur zehn Einzelversuchen und jeden Tag ein neues Problem. In diesen wenigen Einzelversuchen ist ein Tier zu Beginn des Testes nicht fähig, ein Einzelproblem zu lösen. Je weiter aber der Test geht, um so weniger Fehler macht ein Tier pro neues Problem. Als Erklärung muß angenommen werden, daß das Tier an die neuen Probleme unter Verwendung der früher gemachten Erfahrungen bzw. der früher gemachten Fehler geht. Bildlich gesprochen, wird also verfolgt, wie ein Affe „lernt zu lernen", wieweit er aus früheren Situationen auf neue „abstrahieren" kann. Daß diese Leistung höherer „symbolischer" oder „komplexer" Natur ist, vermag besser als theoretische Überlegungen ein phylogenetischer Hinweis zu erläutern. Fische vollbringen diese Leistung nicht, Ratten nur sehr beschränkt. Niedere Affen und Katzen zeigen bemerkenswerte „Abstraktionsleistungen", Schimpansen bringen es zu „affenartiger Behendigkeit" und der Mensch zeigt hier seine höchsten Leistungen. Ähnlich steht es mit den anderen erwähnten symbolischen Prozessen.

II. Erlernen von visuellen Assoziationen

Im Jahre 1936 erwähnten Klüver und Bucy (45) als Folgen der Entfernung des Temporallappens beim Macacus-Affen neben erhöhter oraler und allgemeiner Explorationstendenz, verminderter emotioneller Erregbarkeit und Hypersexualität auch „Seelenblindheit". Im Hinblick auf den Befund der „Seelenblindheit" stellten sich der weiteren Forschung die folgenden Fragen: Besteht innerhalb des Temporallappens des Macacus-Affen ein kritisches fokales Feld für höhere visuelle Funktionen, das lokal von anderen Feldern abgrenzbar ist, die für den übrigen Teil des von Klüver und Bucy erwähnten Symptomenkomplex verantwortlich sind? Sind außerhalb des Temporallappens keine anderen sekundären Cortexfelder mit höheren assoziativen visuellen Funktionen verbunden? Wie läßt sich das mit „Seelenblindheit" umschriebene Verhaltensdefizit experimentell genauer charakterisieren? Werden außerhalb der visuellen psychischen Funktion keine anderen höheren psychischen Funktionen durch die Ausschaltung des Temporallappens betroffen? In welcher funktionellen Beziehung stehen die in Frage stehenden visuellen Felder des Temporallappens zu dem primären optischen System? Eine Reihe von Arbeiten der folgenden Jahre vermochte diese Probleme dem Verständnis näher zu bringen.

Fokalgebiet visueller Lernfunktion im Temporallappen. PRIBRAM und BAGSHAW (88) untersuchten Affen nach Teilausschaltungen des Temporallappens systematisch auf den von KLÜVER und BUCY beschriebenen Symptomenkomplex. Die Autoren fanden, daß nach der Entfernung des frontotemporalen Cortex (Gyrus orbitalis posterior, anteriorer insularer und periamygdaler Cortex) kein Defizit der Sehkraft und der Sehschärfe auftrat. Diese Tiere waren auch fähig, visuelle Assoziationen wie normale Tiere zu erlernen, ebenso konnten sie Tastreize erkennen und ihre spontane Aktivität blieb unverändert. Dagegen traten bei diesen Tieren Veränderungen des Grundumsatzes auf. Auch fraßen sie ohne sichtliche Abneigung Futter, dem Chinin zugegeben war und verhielten sich gegenüber aversiven und sozialen Reizen weitgehend indifferent. Auf der anderen Seite führte die Ausschaltung des lateralen temporalen Cortex (Area 20, 21, 22 der Abb. 2) selektiv zu einem stark verlangsamten Erlernen visueller Assoziationen. Somit ist eine funktionelle Differenzierung des Temporallappens nachgewiesen, indem der temporale laterale sekundäre Neocortex mit der Fähigkeit in Beziehung gebracht werden kann, visuelle Assoziationen zu erlernen, während der übrige Teil des von KLÜVER und BUCY beschriebenen Symptomenkomplexes durch Ausschaltung des Allocortex und des Juxtaallocortex des Temporallappens verursacht wird. Den für die visuelle Funktion kritischen Teil des temporalen Isocortex haben später MISHKIN und PRIBRAM (64) weiter unterteilt. Sie exstirpierten bei zwei Pavianen den ganzen Temporallappen, bei zwei weiteren Tieren den superioren lateralen Teil des temporalen Isocortex und bei den vier übrigen Tieren der Studie die ventrolaterale isocorticale Rindenfläche, inklusive dem Hippocampus. Es zeigte sich, daß sowohl die lobektomierten Tiere wie die Tiere mit der ventrolateralen Resektion des Cortex die visuellen Assoziationen viel langsamer erlernten, als die beiden Tiere, bei denen der superiore temporale Neocortex entfernt worden war. MISHKIN (65) unternahm eine noch weitere Einengung dieses kritischen ventrolateralen Cortex. Er entfernte bei drei Macacen den ventrolateralen Cortex des Temporallappens unter strenger Schonung der tiefergelegenen Strukturen des Hirnlappens. Bei drei Tieren resezierte er durch eine Incision die tiefer gelegene Hippocampusformation. Bei zwei Kontrolltieren nahm er nur die Incision vor, die bei der zweiten Gruppe von

Tieren notwendig war, um die Hippocampusformation zu entfernen. Der Autor fand das Vollbild des visuellen Lerndefizits nur bei den Tieren mit der ventro-lateralen Rindenausschaltung, nicht bei den Affen mit der Hippocampus-resektion. Damit ergibt sich aus dieser Reihe von Studien, daß innerhalb des Temporallappens tatsächlich ein circumscriptes Feld vorhanden ist, welches für die Erwerbung visueller Assoziationen von essentieller Bedeutung ist. Die nach einer Darstellung von Pribram und Mishkin (91) angefertigte Abb. 4 vermittelt einen Überblick über den Umfang dieses Fokalgebietes. Die Abbildung enthält neben der Rekonstruktion der Ausschaltung auch Quer-schnitte durch die Ausschaltung und das Gebiet der retrograden Degeneration im Thalamus. Wie die Abbildung zeigt, ist der Umfang dieses Fokalgebietes

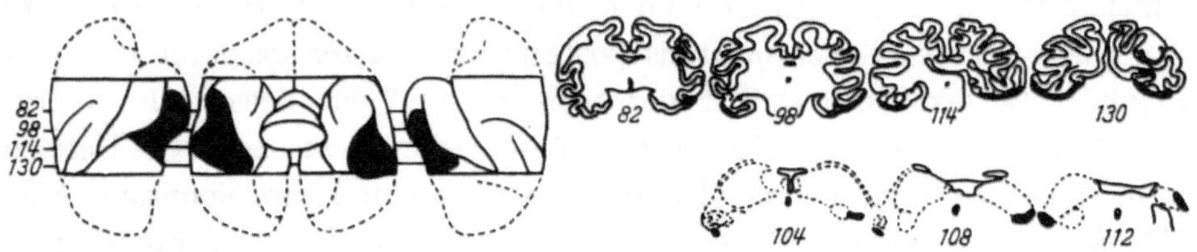

Abb. 4. Fokalgebiet für visuelle Lernleistungen im Temporallappen nach Pribram und Mishkin (91)

für das Erwerben von visuellen Assoziationen identisch mit der früher schon in der Abb. 2 dargestellten Standard-„Temporalausschaltung". Neben diesen Studien führten auch die Untersuchungen anderer Autoren wie Chow (13), Wilson (122), Pribram und Barry (85) und Pasik et al. (83) unabhängig von-einander zum gleichen Ergebnis.

Sekundärer Cortex außerhalb des Temporallappens und assoziative visuelle Funktionen. In einer großen Zahl von Einzelstudien wurde das visuelle Lern-vermögen von Macacen nach der Entfernung von sekundärem Cortex außer-halb des Temporallappens untersucht. Alle diese Arbeiten stimmen darin überein, daß von solchen Ausschaltungen kein visuelles Lerndefizit resultiert. Erwähnt seien von den verschiedenen Arbeiten dieser Gruppe hier nur jene von Blum et al. (6), Jacobsen (39) und Lashley (47). Es muß jedoch bei-gefügt werden, daß alle diese Autoren die visuelle Lernfähigkeit von Affen nach dem Prinzip der Simultanwahl testeten. Die Ergebnisse einiger Arbei-ten mit visuellem Serielernen oder mit sukzessiven Unterscheidungswahlen ergaben ein abweichendes Resultat. Da solche widersprechende Ergebnisse nur nach Ausschaltung des frontalen Assoziationscortex auftraten, sollen diese Studien später bei der Besprechung der Funktionen des frontalen Cortex diskutiert werden.

Nichtvisuelle Lernleistungen und temporaler Cortex. Die schon erwähnten Studien von Pribram und Bagshaw (88), Mishkin und Pribram (64) und von Mishkin (65) befaßten sich auch mit der Lernfähigkeit ihrer operierten Tiere für die aufgeschobene Reaktion. Keiner der Autoren fand bei seinen temporal operierten Macacen ein Lerndefizit für aufgeschobene Reaktionen. Auch aus vielen anderen Arbeiten geht die gleiche Schlußfolgerung hervor, daß die Fähigkeit der aufgeschobenen Reaktion durch die Ausschaltung des temporalen Cortex nicht beeinträchtigt wird. Weiter wiesen PASIK et al. (83), WILSON (122) und PRIBRAM und BARRY (85) nach, daß Ausschaltung des temporalen Cortex die Unterscheidung verschiedener Tastreize nicht beeinträchtigt. WEISKRANTZ und MISHKIN (120) wiesen zudem nach, daß temporal operierte Affen ihre Fähigkeit, auditorische Reize unterscheiden zu lernen, nicht verlieren. Weniger eindeutig waren die Ergebnisse, wenn Tiere auf ihre Lernfähigkeit gustatorische oder Geruchsreize zu unterscheiden getestet wurden, nachdem Ausschaltungen im Gebiete des Temporallappens gesetzt worden waren. Diese Ergebnisse sollen später im entsprechenden Zusammenhang besprochen werden.

Art und Ausmaß des visuellen Lerndefizits nach Beschädigung des Temporallappens. Es ergab sich aus verschiedenen Arbeiten, daß die Entfernung des inferotemporalen Neocortex ein verschieden großes Lerndefizit erbrachte, je nachdem, unter welchen Bedingungen die Tiere getestet wurden.

Ob längere Zeit nach der Operation der Verhaltensschaden von temporal operierten Tieren spontan verschwinde, und ob das quantitative Ausmaß visuellen Erlebens während der Erholungszeit nach der Operation die Größe des Verhaltensdefizites beeinflusse, wurde von CHOW (14) in den Mittelpunkt einer Studie gestellt. Seine fünf Affen erhielten temporale Cortexausschaltungen nachdem sie visuelle Aufgaben erlernt hatten. Ihr Gedächtnis für diese Assoziationen wurde zum Teil 12 Tage nach der Operation, zum Teil aber erst 3 Monate nach der Operation gemessen. Diese Erholungszeiten verbrachten einzelne Tiere in normaler Umgebung und andere Tiere in vollkommener Dunkelheit. Aus dem Beobachtungsmaterial der Arbeit lassen sich Hinweise erhalten, daß eine geringe Wiedererholung verlorener Assoziationen spontan als Funktion der Länge der Erholungszeit eintrete und daß eine solche Wiedererholung nicht eintrete, wenn die Erholungszeit in vollkommener Dunkelheit verbracht worden war. Doch ist das Material der Arbeit zu gering und die Ergebnisse zu wenig eindeutig, als daß der Befund als statistisch gesicherte Allgemeingültigkeit akzeptiert werden könnte.

Daß eine temporale Rindenausschaltung nicht einfach einen Gedächtnisverlust verursacht, zeigten RIOPELLE und CHURUKIAN (96). Ihre operierten Affen lernten nämlich nicht langsamer, wenn die Einzelversuche eines Lerntestes zeitlich weit auseinander lagen, als wenn die Einzelversuche gehäuft erfolgten.

CHOW und ORBACH (17) und ORBACH und FANTZ (80) konstatierten, daß eine visuelle Assoziation nach Entfernung des temporalen Neocortex nicht verloren geht, wenn diese Assoziation vor der Operation „übertrainiert" wurde. MISHKIN und PRIBRAM (64) fanden in ihrer bereits zitierten Arbeit, daß das Lerndefizit der operierten Tiere für das Neuerlernen einer Aufgabe größer war, als für das Wiedererlernen einer bereits früher erlernten Aufgabe. Ferner fanden die beiden Autoren auch, daß das Lerndefizit um so stärker und deutlicher in Erscheinung trat, je schwerer eine visuelle Unterscheidung an und für sich war. Dieser Befund ist konstant aus den Arbeiten aller Autoren zu erheben, die temporal operierte Affen auf ihre visuelle Lernfähigkeit untersuchten. So fand Chow (15), daß die Unterscheidung vieler visueller Muster im Serielerntest nach der Operation vollkommen verloren ging, während die Unterscheidung einer großen Zahl von dreidimensionalen Gegenständen, die ebenfalls mit dem Serielerntest vor der Operation erlernt wurde, nach dem operativen Eingriff nur teilweise verloren ging, und im Gegensatz zum schwereren Musterunterscheidungstest relativ rasch wieder erlernt werden konnte. Hier liegt der Grad der Schwierigkeit in der Unterscheidbarkeit der Discriminanda. Affen erlernen viel rascher, Gegenstände oder Farben zu unterscheiden als schwarz-weiße Muster. Mishkin und Hall (66) trainierten Affen, Größenunterschiede zu unterscheiden. Sie fanden auch hier, daß die operierten Tiere um so mehr Mühe hatten, zu einem Lernerfolg zu gelangen, je kleiner der Größenunterschied der Discriminanda war. Riopelle und Ades (95) kamen zu einem analogen Ergebnis. In einer weiteren Arbeit variierten Pribram und Mishkin (91) den Schwierigkeitsgrad nicht als Funktion der Unterscheidbarkeit der Discriminanda, sondern ließen die Tiere lernen, ein und dieselben Objekte auf eine einfache oder sehr schwierige Art zu unterscheiden. In einem ersten Test benützten sie die einfache Simultanwahl für die Unterscheidung einer Tabakdose von einer Tabakpfeife. Die beiden Gegenstände lagen direkt auf den Futterbehältern, unter der Pfeife immer die Belohnung, und diese bald auf der linken, bald auf der rechten Futterkiste. In einem zweiten Test gelangte die Sukzessivwahl zur Anwendung. Dabei legte man auf den einzigen Futterbehälter des WGTA entweder die Tabakpfeife mit einer Belohnung in dem Behälter, oder die Tabakdose ohne eine Belohnung. Das Tier hatte also zu lernen, den Behälter nur zu öffnen, wenn die Pfeife auf dem Behälter lag. In einem dritten Test nach dem Schema der „Bedingungswahl" wiederum lag nur einer der beiden Gegenstände in der Mitte des Wahlbrettes, jedoch waren zwei Futterbehälter vorhanden, einer links und einer rechts vom Wahlgegenstand. Lag die Dose in der Mitte, so konnte die Belohnung nur im rechten Behälter gefunden werden, lag die Pfeife in der Mitte, so konnte dagegen die Belohnung immer nur in dem linken Futterbehälter gefunden werden. Die nicht operierten Tiere erlernten die erste Bedingung in etwa 1 Tag, die zweite Bedingung in etwa 3 Tagen und die letzte Bedingung in etwa 16 Tagen zu je 30 Einzelversuchen. Die temporal operierten Tiere brauchten kaum doppelt so viele Versuche, um die erste, nämlich die leichteste Testvariation, zu erlernen. Für die beiden schwereren Testvariationen dagegen benötigten sie etwa 3—6mal so viel Versuche wie die unoperierten Kontrolltiere um zum Lernerfolg zu gelangen. Dieses Ergebnis ist sehr interessant; es sagt aus, daß operierte Tiere nicht nur dann versagen, wenn der visuelle Unterschied zwischen zwei Wahlgegenständen oder Wahlmustern zu gering ist, sondern besonders auch dann, wenn der Zusammenhang zwischen einem beliebigen Discriminandum und Futter schwieriger zu erkennen ist. Das Verhaltensdefizit der Tiere ist also weniger „perceptiver" als vielmehr „integrativer" Art. Die gleiche Folgerung läßt sich aus der Arbeit von CHOW und ORBACH (17) ziehen. Hier wurde die Perception der Reize erschwert, indem diese nur für die kurze Zeit von 20 ms bis 1 sec auf die Futterbehälter aufprojiziert wurde. Die Unterscheidung dieser kurzzeitig projizierten Reize wurde vor der Operation erlernt und anschließend übertrainiert. Nach der Operation konnten die Tiere die Reize auch noch unterscheiden, trotzdem sie nur so kurzzeitig sichtbar waren. Daß sie die Reize überhaupt noch sinngemäß unterscheiden konnten, muß, wie schon erwähnt, dem Übertraining zugeschrieben werden. Für perceptive und integrative visuelle Funktionen scheint der Cortex des Affen demnach funktionell differenziert zu sein, wie auch aus dem Vergleich früherer Arbeiten geschlossen werden kann. So fanden HARLOW (31) nach subtotaler Entfernung des Cortex striatus (Feld 17 der Abb. 2) des Macacus-Affen Scotomata und SPENCE und FULTON (106) eine verminderte Sehschärfe, während KLÜVER (44) und SETTLAGE (105) auf der anderen Seite nach ähnlichen Ausschaltungen keine verminderte visuelle Lernfähigkeit fanden. WILSON und MISHKIN (123) unternahmen die Aufgabe, die Wirkung von Ausschaltungen der Area striata und des temporalen Neocortex auf perceptive visuelle Leistungen einerseits und integrative visuelle Lernleistungen andererseits direkt miteinander zu vergleichen. Die Autoren entfernten bei drei Affen den inferotemporalen Cortex, bei drei weiteren die ganze laterale Fläche der Area striata ohne deren medialen Anteil und drei weitere Tiere verwendeten sie als unoperierte Kontrolltiere. Nach TALBOT und MARSHALL (110) erhält der laterale Teil der Area striata beim Affen die Projektionen von etwa 9⁰ des zentralen Sehfeldes. WILSON und MISHKIN testeten alle Tiere auf ihre Sehschärfe, Futtererkennung, Sehfeld, richtige Erkennung von gekreuzt hingelegten Fäden, an denen Futter in die Nähe gezogen werden konnte, Erlernung solche Fäden korrekt zu ziehen, wenn sie mehrfach gekreuzt ausgelegt wurden, ferner die Erlernung von Objektunterscheidung und Musterunterscheidung und letztlich auf die Fähigkeit auf neue Bedeutungen von erlernten Mustern umzulernen. Bei den Tieren, deren lateraler Anteil der Area striata entfernt wurde, fiel auf, daß sie im Gegensatz zu den temporal operierten Tieren mehr Mühe hatten, ein größeres von einem kleineren Objekt zu erkennen, wie sie es vor der Operation gelernt hatten, sofern der Größenunterschied zwischen den beiden Wahlobjekten verkleinert wurde. Ferner war ihre Fähigkeit zu erkennen, an welchem von zwei Fäden Futter befestigt war, geringer als die der temporal operierten

Tiere, sobald die Fäden gekreuzt wurden. Auf der anderen Seite hatten die temporal operierten Tiere deutlich mehr Mühe, die Bedeutung verschiedener Objekte zu erlernen oder die Bedeutung verschiedener visueller Muster zu erfassen. Zu diesem Resultat muß jedoch der wichtige Umstand hinzugefügt werden, daß die nicht operierten Tiere in jedem dieser vier Teste immer noch besser waren als beide operierten Gruppen. Daraus läßt sich schließen, daß Ausschaltungen des Cortex striatus nicht ausschließlich, sondern vorwiegend perceptive Leistungen beeinträchtigen, und daß analogerweise Ausschaltungen des temporalen Cortex vorwiegend integrative Leistungen beeinträchtigen. Das Attribut „vorwiegend" stellt die Ergebnisse in Kontrast zu den früheren schon erwähnten Arbeiten, die Grund zur Annahme boten, daß der Cortex striatus nur eine perceptorische und der temporale Cortex eine nur integrative visuelle Funktion besitze. Aus den Ergebnissen von Wilson und Mishkin liegt es dagegen nahe zu schließen, daß die beiden neuralen Substrate voneinander eng abhängen und ihre funktionellen Aufgaben sich im Rahmen einer begrenzten Spezialisierung ergänzen.

Die Gesamtheit der Arbeiten über Art und Ausmaß des visuellen Lerndefizites nach Beschädigung des temporalen Cortex berechtigt somit zu einer Reihe von Schlüssen: — Am Erwerben einer neuen assoziativen Leistung ist der inferotemporale Cortex beim Macacus-Affen in kritischer Weise beteiligt. — Je stärker eine solche Assoziation im Verlaufe des Training gefestigt oder gar „übertrainiert" wurde, um so mehr wird eine solche Leistung von der Intaktheit des inferotemporalen Cortex unabhängig. — Wenn eine neue Assoziation leicht erlernbar ist, kann sie auch erworben werden, ohne daß der temporale Cortex vorhanden ist, jedoch nur unter größerem Lernaufwand, als von einem Normaltier. — Andere Hirnstrukturen können also für den spezifischen inferotemporalen Cortex einspringen; welches aber diese Strukturen sind, wurde bis jetzt nicht erforscht. — Wenn auch die Funktion des inferotemporalen Cortex weitgehend integrativer und jene des Cortex striatus weitgehend perceptiver Natur ist, darf doch diese funktionelle Trennung nicht als absolut angenommen werden. Es bestehen vielmehr zwischen den beiden Gebieten funktionelle Wechselwirkungen.

Funktionelle Beziehungen zwischen temporalem Cortex und primärem visuellen System. Einhellig berichten die Autoren, die temporalen Cortex beim Macacus-Affen resezierten, über retrograde Degenerationen im Pulvinar. Diese Beziehungen zwischen Pulvinar und temporalem Cortex können auch mit elektrophysiologischen Methoden demonstriert werden, wie Jasper et al. (41) und auch Niemer et al. (78) zeigten. Es lag daher nahe, zu untersuchen, ob die Zerstörung dieses Thalamuskernes einen ähnlichen Effekt nach sich ziehen könnte, wie die Resektion des temporalen Cortex selber. Chow (16) beschritt diesen Weg, doch konnte er keinen visuellen Lernverlust nach Elektrokoagulation des Pulvinars nachweisen. Whitlock und Nauta (121) wiesen auf Verbindungen vom temporalen Neocortex zu den Colliculi superiores hin, die in diesem Zusammenhang ebenfalls von Bedeutung sein könnten. Doch fanden ROSVOLD et al. (99) nach Elektrokoagulation dieses Kerngebietes ebenfalls kein visuelles Lerndefizit beim Affen. ETTLINGER (22) widmete eine eingehende Studie der nach diesen negativen Ergebnissen verbleibenden Alternative, nämlich ob die funktionell wichtigen Afferenzen dem temporalen Cortex aus der primären Sehbahn (von den Corp. geniculata lat. nach dem Cortex striatus) oder aus der primären Sehrinde (Cortex striatus) zugeleitet werden. Die Tatsache, daß zwar zwischen den beiden Sehrinden quere Verbindungen durch das Corpus callosum bestehen, nicht aber zwischen den beiden Sehbahnen, benützte ETTLINGER zu einem geschickten Kombinationsspiel mit serienweise ausgeführten Teilausschaltungen. Vorerst zeigte der Autor, daß einseitige Durchtrennung des linken Tractus opticus (unmittelbar hinter dem Chiasma opticum) nicht zum Verlust einer früher erlernten Assoziation führt, auch dann nicht, wenn der Eingriff von der Durchtrennung des Corpus callosum begleitet oder gefolgt wird oder von der Resektion des gleichseitigen linken temporalen Cortex. In allen drei Fällen verbleibt der rechtseitige temporale Neocortex intakt, und kann Afferenzen entweder aus der primären gleichseitigen Sehrinde oder aus der gleichseitigen Sehbahn erhalten. Daß ein einziger intakter temporaler Lappen zum visuellen Lernen genügt, zeigten früher schon MISHKIN und PRIBRAM (64). Wird andererseits nach Sektion der linken Sehbahn der rechte, also gegenseitige temporale Cortex entfernt, so entsteht ein beträchtlicher Verlust der Assoziation. Das heißt, daß jeder Temporallappen seine visuellen Afferenzen vorwiegend aus Strukturen der gleichseitigen Hirnhälfte bezieht, sei es aus Sehbahn oder Sehrinde. Nach Zerschneidung des linken Traktes und Resektion des rechten temporalen Cortex verbleibt der linke temporale Cortex, der seine Afferenzen nun nur noch aus dem gegenseitigen rechten Cortex striatus erhalten kann. Um eine solche funktionelle Afferenz zu beweisen, zerstörte ETTLINGER den noch verbleibenden linken temporalen Cortex oder durchschnitt das Corpus callosum in einer weiteren Sitzung. Durch den Eingriff trat zum schon vorher bestehenden schweren assoziativen Defizit noch eine zusätzliche drastische Verstärkung des Defizites hinzu. Daraus läßt sich mit Sicherheit schließen, daß funktionelle Afferenzen aus der primären Sehrinde in den temporalen Cortex gelangen. Aus einer Arbeit von MISHKIN (70), der mit der Methode serienweiser Ausschaltungen auf ähnliche Weise vorging, kann der gleiche Schluß gezogen werden, nämlich, daß funktionelle Afferenzen den temporalen Cortex aus der primären Sehrinde erreichen müssen. Ob dies durch direkte transcorticale Verbindungen oder subcorticale Assoziationsbahnen der Fall ist, läßt sich bis heute nicht entscheiden. Zwar fanden CHOW (13) und auch LASHLEY (47) nach der Resektion der zwischen temporalem Cortex und primärer Sehrinde gelegenen Area praestriata (Area 18 und 19 der Abb. 2) keinen Verlust erlernter Assoziationen, der auf direkte transcorticale Verbindungen zwischen den beiden Rindengebieten schließen ließe. Jedoch waren die von beiden Autoren gesetzten Ausschaltungen der Area praestriata nicht mit Sicherheit genügend vollständig. Eine weitere Gruppe von Autoren widmete sich dem Studium der Funktion des Corpus callosum für visuelle Assoziationen. Ihre Technik bestand darin, bei Versuchstieren das Chiasma opticum in mediosagittaler Ebene zu durchtrennen, ein Auge des Tieres zu verdecken, und das Tier nur mit einem Auge lernen zu lassen. Später wurde getestet, ob das Tier die Aufgabe auch mit dem anderen beim Lernen zugedeckten Auge erfassen konnte. Myers (73, 74) fand so, daß eine durch isolierte Afferenz in die eine Hirnhälfte erworbene Assoziation auch auf die andere Hirnhälfte unmittelbar übertragen wird, wenn das Corpus callosum noch intakt ist, nicht aber wenn auch dieses durchschnitten wurde. Ferner zeigte Myers (75) daß für die Übertragung einer erlernten Assoziation von der einen in die andere Hirnhälfte die hinteren 25 % des Corpus callosum genügen. Sperry et al. (108) demonstrierten sogar, wie nach Durchtrennung von Chiasma und Corpus callosum beide Hirnhälften weitgehend unabhängig voneinander verschiedene visuelle Assoziationen erwerben können. Obschon diese Arbeiten über die Funktion des Corpus callosum nicht den Affen, sondern die Katze als Versuchstier benützten, stellen sie doch eine interessante Ergänzung zu den Befunden am Macacus dar. In bezug auf die Frage, wie die eigentliche „Mechanik der Integration" von Sehreizen im Cortex vor sich gehe, folgt die moderne „Köhlersche Feldtheorie" wie ein roter Faden den alten Hypothesen Pavlovs. Pavlov glaubte, daß ein Reiz von einem Punkt ausgehend über den ganzen Cortex „irradiere" und sich anschließend auf einen kritischen Punkt „konzentriere". Köhler und Held (46) zeigten direkt, wie langsame Potentialänderungen auf dem Cortex parallel gehen zur Bewegung eines Reizes, der quer durch das Gesichtsfeld eines Tieres geführt wird. Doch gelang es weder Lashley et al. (48) mit kreuz und quer in den Cortex gespickten Goldplättchen, noch Sperry et al. (107) mit gleicherweise applizierten Tantalplättchen noch Thomas und Stewart (111) durch Durchströmen des Cortex mit Gleichstrom in der einen oder anderen Richtung irgend welche nachteilige Wirkungen auf visuelles Verhalten und visuelle Assoziationen zu erzeugen. Damit wird eine alte, und in anderen Formen immer wieder postulierte Theorie über den Mechanismus des Aufbaus von Assoziationen im Cortex erneut in Frage gestellt.

Die in diesem Kapitel besprochenen Arbeiten zeigen, wie beide Hirnhälften im Aufbau von Assoziationen zusammenarbeiten, obwohl jede Hemisphäre auch für sich allein visuelle Zusammenhänge erlernen kann. Ferner ist es sehr wahrscheinlich, daß jeder temporale Cortex seine Informationen mit Vorzug aus der gleichseitigen primären Sehrinde und zum kleineren Teil aus der Sehrinde der Gegenseite bezieht, und daß der Austausch zwischen den beiden Hemisphären vom hinteren Viertel des Corpus callosum bewältigt wird.

III. Erlernen von Gehörassoziationen

Über die Fähigkeit verschiedene auditorische Reize zu verassoziieren, bestehen bis heute beim Macacus-Affen sehr wenig experimentelle Arbeiten. Dies hat seinen Grund darin, daß der Affe im Gegensatz zu anderen Unterscheidungen die Bedeutung von auditorischen Signalen nur außerordentlich langsam erlernt. Alle bisherigen Arbeiten benötigten z.B. für ein Einzeltier im WGTA mehr als 1000 Einzelversuche bis zum eindeutigen Erwerben einer auditorischen Assoziation, sofern nur Futtermotivierung, und nicht auch Strafmotivierung verwendet wurde. Das bedeutet also etwa 2 Monate lang täglich eine Versuchssitzung zu 20 Einzelversuchen und etwa 20 min Dauer.

Interessant sind dagegen eine Reihe von Arbeiten, die NEFF mit seinen Mitarbeitern (12, 20, 29, 76, 77) an Katzen mit ähnlichen Techniken ausführte, wie sie auch beim Affen geläufig sind. Diese Autoren fanden vorerst, daß die Entfernung der primären Hörrinde die Katze nicht daran hinderte, die Unterscheidung zwischen verschiedenen Tonfrequenzen zu erlernen. Der Umfang der Ausschaltung wurde dabei so gestaltet, daß alles Gebiet erfaßt wurde, das mit elektrophysiologischen Methoden als auditorischer Cortex definiert ist. Weiter fanden diese Autoren, daß sogar eine nach elektrophysiologischen Maßstäben wie nach dem Kriterium der retrograden Degeneration der Corp. geniculata medialia vollständige Entfernung des auditorischen Cortex die Fähigkeit der Katze, verschiedene Tonfrequenzen zu unterscheiden, nicht beeinträchtigt. Jedoch führt eine solch komplette Entfernung des primären auditorischen Cortex dazu, daß eine Katze nicht mehr lokalisieren kann, aus welcher Richtung des Raumes sie eben einen Ton gehört hat. In einer weiteren Studie erweiterten diese Forscher das Ausmaß der Ausschaltungen nach ventral, so daß sie das Gebiet miterfaßten, das beim Macacus-Affen für visuelles Lernen unerläßlich ist, nämlich die Felder 20 und 21 des inferotemporalen Cortex. Bei so erweiterten Ausschaltungen fanden sie, daß die Katzen die Fähigkeit verloren, auditorische Klangmuster (= Tonfolgen) voneinander zu unterscheiden, während sie die Fähigkeit Klangfrequenzen zu unterscheiden immer noch weiter behielten. Dieser Verlust trat nicht auf, wenn von der Ausschaltung des eigentlichen primären auditorischen Cortex einzelne Teile ausgespart wurden. Somit entsteht der interessante Befund, daß für die Unterscheidung von Klangmustern bei der Katze der gleiche Cortex beteiligt ist, der beim Affen wichtig ist für die visuelle Unterscheidung. Ob hier ein Widerspruch zu den Ergebnissen beim Affen liegt, kann heute noch nicht mit Sicherheit gesagt werden. Ob der inferotemporale Cortex bei der Katze eine ebenso kritische Bedeutung für visuelles Unterscheidungslernen hat, wie beim Affen, läßt sich ebenfalls nicht sagen, da entsprechende Experimente an der Katze fehlen. Wie bereits erwähnt, entfernten WEISKRANTZ und MISHKIN (120) bei einigen Affen den inferotemporalen Cortex allein und fanden keine Beeinträchtigung der Fähigkeit der auditorischen Unterscheidung, doch testeten

Weiskrantz und Mishkin die Affen nicht auf die Unterscheidung zwischen Klangmustern, sondern zwischen einem weißen Rauschen und einer bestimmten Klangfrequenz. Analoge Verhältnisse zur Katze ergäben sich also erst, wenn gezeigt würde, daß inferotemporal operierte Affen mit zugleich beschädigtem auditorischem Cortex komplexe Klangmuster nicht unterscheiden können. In diesem Zusammenhang ist ein Hinweis auf die übrigen experimentellen Gruppen der Studie von Weiskrantz und Mishkin interessant. Die Autoren entfernten bei einer zusätzlichen Gruppe von Tieren den „posterioren temporalen Cortex" wobei zwar die primäre Hörrinde gemäß anatomischer Definition auf dem Planum supratemporale geschont, jedoch der ganze angrenzende Cortex von Temporal, Parietal und Occipitallappen mitentfernt wurde. Diese Ausschaltung bewirkte einen „geringen" Ausfall der auditorischen Lernfähigkeit, der jedoch nicht signifikant war. Bei einer zusätzlichen Affengruppe resezierten die beiden Autoren den lateralen frontalen Cortex. Dabei fanden sie ein ausgesprochenes auditorisches Lerndefizit, und stellten deshalb die Frage, ob beim Affen ein sekundäres auditorisches Hirnrindenfeld entgegen der Erwartung nicht in der parieto-occipito-Temporalgegend, sondern im frontalen Cortex zu suchen sei. Der Befund eines auditorischen Lerndefizites nach Beschädigung des frontalen Cortex findet jedoch in einigen anderen Arbeiten keine Bestätigung, und es stellen sich vielmehr Zweifel, ob dieses Defizit überhaupt sensorischer Natur sei oder nicht. Aus diesem Grunde soll die Frage einer auditorischen Funktion des frontalen Cortex in einem späteren Zusammenhang näher besprochen werden.

Die Ergebnisse der in diesem Kapitel besprochenen Arbeiten müssen also dahin zusammengefaßt werden, daß feste Schlüsse weder in bezug auf den Affen noch in bezug auf die Katze gezogen werden können. Erst weitere Arbeiten können zeigen, ob „sekundäre oder höhere" auditorische Rindenfelder lokalisiert werden können, und ob diese eventuell, wie die Arbeiten an der Katze andeuten, in den Temporallappen, eventuell sogar in das gleiche Gebiet wie das höhere visuelle Rindenfeld zu liegen kommen.

IV. Erlernen von Geruchsassoziationen

Über dieses technisch schwierig zu untersuchende Gebiet existieren bis heute ebenfalls nur sehr wenig experimentelle psychophysiologische Untersuchungen. Bagshaw und Pribram (2) beschrieben in einer frühen Arbeit, daß Affen nach Resektion des Temporallappens chininhaltiges Futter ohne Zeichen von Abneigung fraßen. In ihrer ursprünglichen Arbeit über das Temporalhirnsyndrom beim Macacen erwähnten Klüver und Bucy (45), daß lobektomierte Tiere Objekte in auffallender Weise mit Schnüffeln zu untersuchen trachten. Kaada (42) konnte ferner mit elektrophysiologischen Methoden direkte Verbindungen zwischen dem Bulbus olfactorius und dem inferioren Teil des Neocortex des Temporallappens nachweisen. Auf Grund dieser Hinweise unternahmen es Santibanez und Hamuy (103), dem Macacus-Affen Geruchsassoziationen anzutrainieren. Der Test wurde mit dem WGTA ausgeführt. Vor dem Tier, ungefähr auf Kopfhöhe, stand im Wahlraum des WGTA eine vertikal aufgestellte Platte mit zwei kreisrunden Löchern, hinter denen Hängetürchen aus Maschengitter so befestigt waren, daß sie vom Tier weggeschoben werden konnten. Je ein Wattebausch mit Orangenextrakt und ein anderer mit Vanilleextrakt wurden am unteren Ende der beiden Türchen außerhalb der Sicht des Tieres bald links bald rechts befestigt. Nur durch Wegschieben des nach Vanille riechenden Türchens konnte das Tier mit seiner Hand ein Erdnüßchen erreichen. Der ganze Test ist also in jeder Hinsicht analog dem Prinzip der Simultanwahl beim Erlernen der visuellen Unterscheidung. Fünf Affen wurden operiert, und zwar wurde bei zwei Tieren der ventrale Teil des Temporallappens entfernt, wobei subcorticale Strukturen und Juxtaallocortex miterfaßt wurde, während bei den drei anderen Tieren der laterale neocorticale Anteil des Temporallappens reseziert wurde. Alle Tiere erlernten die Aufgabe vor der Operation und wurden nach der Operation auf das Gedächtnis getestet. Es zeigte sich, daß vier Tiere nach der Operation sich nicht nur an die gelernte Assoziation nicht mehr erinnerten, sondern sogar mehr Fehler machten als beim Erstlernen, wobei zwei Tiere die Aufgabe in der gesteckten Zeit überhaupt nicht mehr erlernen konnten. Die Zahl der im Test nach der Operation gemachten Fehler ging weitgehend parallel zum Ausmaß der Beschädigung des inferioren lateralen Neocortex des Temporallappens der einzelnen Tiere. Trotzdem kann aus diesen Versuchen nicht mit Sicherheit geschlossen werden, daß das schon besprochene temporale sekundäre höhere Sehfeld und dieses funktionell analoge Geruchsfeld anatomisch kongruent seien. Denn die Ausschaltungen, die Santibanez und Hamuy setzten, sind ausgedehnter als die in all den visuellen Studien klassischerweise ausgeführten Resektionen. Santibanez und Hamuy entfernten zusätzlicherweise den Pol des Temporallappens und den periamygdaloiden Cortex. Es könnte also sein, daß sich die kleinste effektive Ausschaltung gerade auf diesen polaren und periamygdaloiden Cortex beschränken ließe, womit eine funktionsspezifische Trennung zwischen höherem visuellem und höherem Geruchscortex nachgewiesen wäre. Ob diese Möglichkeit zutrifft, wird erst mit weiteren experimentellen Arbeiten nachgewiesen werden können. Es ergibt sich demnach aus dieser Studie von Santibanez und Hamuy eine interessante Parallele zu den schon oben besprochenen auditorischen Studien von Neff und seinen Mitarbeitern. Sowohl für auditorische als auch für olfactorische höhere assoziative Leistungen ist der temporale Cortex wichtig, und in beiden Fällen ist eine genaue Lokalisation innerhalb dieses Rindengebietes noch nicht festgelegt. Denkt man daran, wie der Pol des Temporallappens in einer Reihe von Arbeiten, über die Pribram und Krüger (89) eine Übersicht gaben, sowohl auf Grund anatomischer wie auf

Grund elektrophysiologischer Untersuchungen zu dem olfactorischen afferenten System in enge Beziehung gebracht werden konnte, so drängt sich geradezu ein Besinnen auf die strategisch zentrale Lage des Temporallappens zwischen den primären Rindenfeldern vom Hörsinn, Geruchssinn und Sehsinn auf. Es könnte in diesem Zusammenhang nicht überraschen, wenn der temporale Cortex höhere assoziative Funktionen aller dieser Sinne und auch des Geschmackssinnes besäße. So fanden Hamuy und Mitarbeiter (30) in einer Studie mit Teilresektionen des temporalen Cortex einen Hinweis in dieser Richtung. Sie entfernten bei zehn Macacen schrittweise den inferioren lateralen, superioren lateralen Cortex und den superioren polaren temporalen Cortex. Sie fanden ein großes visuelles Defizit, wenn der superiore polare Cortex entfernt wurde, also ein Teil des lateralen temporalen Cortex, der in den früher besprochenen Arbeiten von Pribram und Mishkin und Chow, sowie anderen weitgehend aus dem Bereich assoziativer visueller Funktionen ausgeklammert worden ist. Ein ausgeprägtes Geruchsunterscheidungsdefizit fanden sie dagegen nach Resektion des inferioren temporalen Cortex. Aus diesen Hinweisen ergibt sich, daß auch die von vielen Autoren als spezifisch angesehene visuelle Funktion des inferioren Teils des temporalen Cortex einer weiteren Verifizierung bedarf, bevor endgültige Schlüsse gezogen werden dürfen

V. Erlernen von Geschmacksassoziationen

Bisher wurden keine Untersuchungen am Affen unternommen, mit dem Ziel, die Abhängigkeit von Geschmacksassoziationen von einzelnen Rindenfeldern nachzuweisen. Es findet sich in der Literatur auch keine Technik, die benützt wurde, um den Affen Geschmacksunterscheidungen anzutrainieren. Die alte Technik Pavlovs, der in das Maul von Hunden eingeführte Säuren oder Bitterstoffe als bedingte Reize anwandte, ist nicht auf Affen übertragen worden. Es gibt daher in dieser Frage nur die wenigen Hinweise aus den bereits erwähnten Arbeiten von Pribram und Bagshaw (88), nach welchen temporal operierte Tiere einen höheren Schwellenwert für die Akzeptierbarkeit von Chinin in der Nahrung aufweisen.

VI. Erlernen von Tastassoziationen

Wie schon früher erwähnt wurde, wiesen die Arbeiten von Pasik et al. (83) und von Pribram und Barry (85) darauf hin, daß die parietooccipitale Rindengegend wichtig für das Vermögen sei, somaesthetische Unterscheidungen zu erlernen, und daß andererseits der temporale Cortex an solchen Funktionen nicht beteiligt sei. Wilson (122) gelang es, in einer eleganten Arbeit diesen älteren Hinweisen bedeutend mehr Gewicht zu verleihen, und ihre Arbeit sei daher hier kurz besprochen. In der Untersuchung wurden Affen mit reseziertem inferotemporalem Cortex Tieren gegenübergestellt, deren parietooccipitaler Cortex wegoperiert worden war. Die „parietooccipitale"

Ausschaltung war begrenzt durch den Sulcus intraparietalis auf der anterioren und superioren Seite, durch den Sulcus lunatus auf der posterioren Seite und durch den Gyrus temporalis superior in der ventralen Richtung. Die Ausschaltung umfaßte also nach der Abb. 1 das Feld Nr. 7 und die superioren Anteile der Felder Nr. 18, 19 und 22 der Nomenklatur nach BRODMANN und entspricht der in Abb. 2 dargestellten parietalen Standardausschaltung. Alle Tiere wurden mit visuellen und Tastreizen getestet. Für das visuelle Unterscheidungslernen wurden auf die beiden Futterkisten des WGTA Deckel gelegt, auf die mit Leisten die Form eines „L" oder des Spiegelbildes eines „L" aufgenagelt war, oder statt den L-Formen wurden kurze und lange Leisten verwendet. Das Tier hatte zu lernen, immer den Behälter zu öffnen, auf dem es das korrekte „L" sah, oder jenen auf dem die lange Leiste aufgenagelt war. Für die Tastunterscheidung wurden die gleichen Merkmale gebraucht, doch wurden die Tiere in völliger Dunkelheit getestet, so daß sie den Unterschied der beiden Futterbehälterdeckel nur durch Tasten erkennen konnten, also ähnlich wie der Blinde die Blindenschrift liest. Die Tiere erlernten beide Aufgaben wiederum vor der Operation, und wurden nach der Operation auf das Gedächtnis getestet. Es zeigte sich, daß die parietal operierten Tiere nach der Operation die Bedeutung der Leisten auf den Futterdeckeln nicht mehr erfassen konnten, wenn sie sie in der Dunkelheit betasten mußten; dagegen waren sie ohne weiteres fähig, die richtigen Behälter zu wählen, wenn sie in vollem Licht die Situation betrachten konnten. Für die inferotemporal operierten Tiere ergab sich genau das Gegenteil. Sie zogen keinen Vorteil von der Betrachtung der Merkmale, konnten sie jedoch in der Dunkelheit durch Betasten unterscheiden. Als Nebenbefund ergab sich, daß die parietal operierten Tiere für kurze Zeit nach der Operation grobe visuelle Störungen aufwiesen, die bei einigen Tieren bis zur Blindheit für 1—2 Tage führte. Die Autorin glaubt, diese Sehdefekte auf Beschädigungen von Fasern der optischen Radiation zurückführen zu dürfen, die bei der Absaugung von angrenzenden Teilen des Occipitalcortex entstanden. Daß diese parietal operierten Tiere trotz Beschädigung der primären Sehkraft im visuellen Unterscheidungstest keinen Schaden hatten, stützt eindrücklich die früher besprochene Arbeit von WILSON und MISHKIN (123) über den differenzierten Effekt der Ausschaltung des occipitalen Cortex striatus und des inferotemporalen Cortex. So ergibt sich also aus dieser Studie das Idealbild einer doppelten Dissoziation der Ergebnisse: Ausschaltung I produziert das Lerndefizit A, nicht aber das Lerndefizit B, während die Ausschaltung II genau das Umgekehrte tut. In einer weiteren Arbeit zeigten ORBACH und CHOW (81), daß Ausschaltung der primären somatischen Area I, also der Felder 3, 1, 2 nach der Einteilung von BRODMANN, das gleiche Defizit der schlechteren Tastunterscheidung noch viel krasser und irreparabler bewirkt, als die Beschädigung der sekundären somatischen Areale, die in der Arbeit WILSONs entfernt wurden. Somit besteht hier eine gewisse Parallele zu den bereits besprochenen Wechselwirkungen zwischen primärem auditorischem und anliegendem sekundären auditorischen Cortex, wie sie an der Katze nachgewiesen wurde.

VII. Symbolische psychische Prozesse

Im Jahre 1936 berichtete Jacobsen (38), daß frontalhirnoperierte Macacen die Fähigkeit zu aufgeschobenen Reaktionen völlig verloren, daß vor der Operation eventuell provozierte emotionelle Stressreaktionen drastisch abgeschwächt wurden, und daß aber andererseits die Fähigkeit visuelle Unterscheidungen zu erlernen in keiner Weise litt. Auf der anderen Seite beschrieben Kennard et al. (43) in eindrücklicher Weise Hyperaktivität als eine weitere Folge von Frontalhirnausschaltungen. Diese Hyperaktivität äußert sich in einem auffallenden ruhelosen im Kreis Herumwandern der Tiere, ähnlich, wie man es bei Löwen im Zwinger sieht. Auf Grund dieser Befunde bewegte sich die Erforschung des Frontalhirnsyndroms beim Macacen vor allem um die folgenden Punkte: — Können diese einzelnen Funktionen anatomisch getrennten Rindenarealen zugeschrieben werden? — Welches ist die kleinste effektive Ausschaltung die den Verlust der aufgeschobenen Reaktion bewirkt? — Kommt die Funktion der aufgeschobenen Reaktion weiteren außerhalb des Frontalhirns gelegenen Rindenarealen, oder eventuell auch subcorticalen Strukturen zu? — Wie kann die psychologische Bedeutung des Verlustes der aufgeschobenen Reaktion durch Abwandlungen der Testmethodik näher charakterisiert werden? — Führen Frontalhirnausschaltungen außer dem Verlust der aufgeschobenen Reaktion zu keinen anderen Verhaltensdefiziten, besonders was andere symbolische psychische oder assoziative Prozesse betrifft?

Corticales Fokalgebiet der aufgeschobenen Reaktion. In seinen ersten Arbeiten entfernte Jacobsen allen präfrontalen Cortex, der anterior einer Linie „einige Millimeter" vor der vorderen Begrenzungslinie des prämotorischen Cortex (Feld 4 der Abb. 2) lag. Später wies er nach, daß Entfernung des prämotorischen und des motorischen Cortex zu keinem Verlust der aufgeschobenen Reaktion führt (40). Im nächsten Jahrzehnt ausgeführte Arbeiten mit dem Versuch, der Funktion der aufgeschobenen Reaktion innerhalb des präfrontalen agranularen Cortex selber ein engeres Fokalfeld zuzuweisen, führten zu widersprechenden Ergebnissen. Erst spätere Arbeiten vermochten mehr Klarheit in diese Frage zu bringen. Pribram et al. (87) verglichen die Wirkung der Ausschaltung des ventromedialen Cortex bei vier Pavianen und jene der Ausschaltung des dorsolateralen frontalen Cortex bei drei weiteren Pavianen mit dem Verhalten von zwei unoperierten Tieren. Die dorsolaterale Operation umfaßte den lateralen und dorsalen frontalen Cortex rostral und dorsal vom Knie und unteren Ast des Sulcus arcuatus (nach vorne geöffneter U-förmiger Sulcus in Frontallappen auf Abb. 2). Bei der ventromedialen Operation wurde der laterale frontale Cortex vollständig geschont, jedoch die ventrale und mediale Rinde beider Frontallappen weitgehend entfernt. Die Autoren fanden in eindeutiger Weise, daß die Ausschaltung des ventromedialen Cortex mit der Fähigkeit zur aufgeschobenen Reaktion kaum, oder nur in einzelnen Fällen geringfügig interferierte, während die Ausschaltung des dorsolateralen Cortex die gleiche Fähigkeit vollständig zum Erlöschen brachte. Blum (7) unterteilte diesen dorsolateralen Cortex weiter durch kleine Teilausschaltungen bei sechs Macacen und postulierte auf Grund seiner Studie, daß das kritische Feld für die aufgeschobene Reaktion auf das Gebiet beidseits und in der Tiefe des Sulcus principalis des Frontallappens (Sulcus principalis = Längssulcus innerhalb des Sulcus arcuatus auf der dorsolateralen Fläche des Frontallappens, gemäß Abb. 1) eingeengt werden könne. Mishkin (69) konnte den gleichen Befund mit einem größeren Kollektiv von zehn Macacen erhärten.

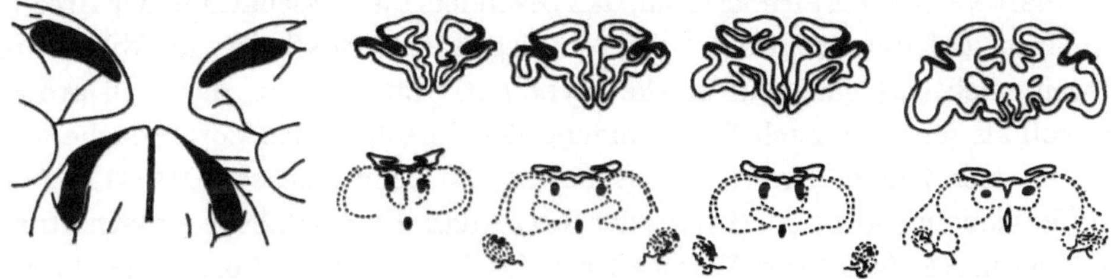

Abb. 5. Fokalgebiet für die Fähigkeit zu aufgeschobenen Reaktionen nach Mishkin (69)

Die Abb. 5, die nach der Arbeit von Mishkin (69) angefertigt wurde, gibt einen Überblick über dieses engere Fokalgebiet für die Fähigkeit zur aufgeschobenen Reaktion. Die Abbildung zeigt neben der Rekonstruktion der Ausschaltung auch Querschnitte der Ausschaltung, sowie Hinweise auf die thalamischen retrograden Degenerationsherde. Besondere Beachtung verdient die vollständige Ausschaltung des Cortex in den Tiefen des Sulcus principalis, wie sie an Hand der Querschnitte festgestellt werden kann.

Inwieweit die frontalen Augenfelder (dorsale Anteile der Area 6 auf der Abb. 2) von diesem Fokalfeld ausgeschlossen werden können, läßt sich nur indirekt aus den bisherigen Studien schließen. Weder Mishkin noch Blum haben in ihren Studien dieses Feld je bei einem Tier separat für sich ausgeschaltet. Pribram (90), der in einer anderen Studie gezielt die frontalen Augenfelder ausschaltete, fand ein starkes Defizit beim Erlernen der aufgeschobenen Reaktion, doch umfaßten seine Ausschaltungen auch einen wesentlichen Teil des Feldes entlang des Sulcus principalis, das von Mishkin und von Blum unabhängig voneinander als Fokalfeld für die aufgeschobenen Reaktionen identifiziert wurde, und das auch nicht mehr zu den frontalen Augenfeldern gezählt werden darf. Somit darf heute zusammenfassend der dorsolaterale Cortex innerhalb des Schenkels des Sulcus arcuatus in den Mittelpunkt der Funktion der aufgeschobenen Reaktion gestellt werden, während andererseits die dem Gebiet caudal anliegenden Augenfelder, wenn nicht mit

Sicherheit, so doch mit großer Wahrscheinlichkeit von der Funktion ausgeschlossen werden können. Damit verbleibt die Frage nach der engeren Lokalisation der beiden anderen erwähnten Hauptkomponenten des Frontalhirnsyndroms beim Macacen, nämlich der Hyperaktivität und der Änderungen des emotionellen Verhaltens. In beiden Fällen lassen sich nicht mit genügender Sicherheit separate Fokalfelder für diese Funktionen von dem Fokalfeld der aufgeschobenen Reaktion abtrennen. In bezug auf die Hyperaktivität fand French (26), daß sie vor allem durch die Resektion der Area 9 des agranularen frontalen Cortex ausgelöst wird, also eines Feldes, in dem auch der kritische Focus für die aufgeschobene Reaktion liegt. Resektion der Area 6, also der Augenfelder bewirkt nach dem gleichen Autor keine Hyperaktivität. Andererseits wiesen verschiedene Autoren, wie Ruch und Shenkin (102) und Livingstone et al. (51) früher nach, daß besonders die Ausschaltung der Area 13 des Macacus-Hirnes (auf der Orbita aufliegende Cortexfläche, in Abb. 2 nur teilweise sichtbar) eine sehr starke Hyperaktivität bewirkt, die noch größer sein soll als jene, die nach Beschädigung der dorsolateralen Cortexfläche auftritt. Ferner fanden Richter und Hines (94) beim Macacen, Mettler (53) beim Pavian und Mettler (54) bei der Katze, daß die zusätzliche Ausschaltung des Nucleus caudatus zur Ausschaltung des dorsolateralen Cortex die Hyperaktivität noch verstärkt. Diese Ergebnisse berechtigen zum Schluß, daß nicht nur große Teile des frontalen Cortex, sondern auch subcorticale Strukturen des Striatums an der Regulation der normalen spontanen Aktivität beteiligt sind. In ähnlicher Weise läßt sich auch für die Veränderung der Emotionalität kein engerer Bezirk innerhalb des Frontallappens angeben. In seinem ersten Bericht beschrieb Jacobsen (39), wie ein Schimpanse vor der Operation „neurotisch" wurde, weil ihm zu schwere Probleme aufgegeben wurden. Je weiter die Teste voranschritten, um so aufgeregter benahm sich das Tier, jeder Fehler den es machte, versetzte es in tiefste Wut mit Bodenstampfen, Wimmern und Schreien, wobei es auch zu häufigen Defäkationen und Urinieren kam. Nach beidseitiger präfrontaler Lobektomie war es, wie der Autor schreibt, „als hätte das Tier alle seine Bürde dem Herrn überantwortet", von nun an reagierte es auf jede schwere Testsituation und gar auf jeden Fehler mit gleichgültiger Wurstigkeit. Seit dieser Arbeit treffen wir in der Literatur auf keine weiteren systematischen Studien über die Wirkung von Frontalhirnausschaltungen auf „experimentelle Neurosen", obwohl es gerade der Bericht von Jacobsen war, der Edgas Moniz dazu veranlaßte, die frontale Lobotomie in die Neurochirurgie einzuführen. Wohl wurden weitere andere psychoexperimentelle Techniken angewandt, wie z.B. die antrainierte Furcht vor einem Gegenstand oder einer Situation, oder die bedingte Flucht auf Schmerz signalisierende Reize. Doch zeigten diese Studien, daß verschiedenartige Veränderungen der Emotionalität inklusive Abschwächungen, wie Jacobsen sie am Schimpansen nach frontaler Lobektomie beobachtete, auch auftreten, wenn eine Reihe von anderen, außerhalb des Frontalhirns liegenden Strukturen entfernt werden. Außerdem fanden HARLOW et al. (34), im Gegensatz zu JACOBSEN, nach Entfernung des frontalen Cortex keine Abnahme der spontanen Emotionalität des Affen bei Mißerfolgen in der Testsituation. Eine Reihe von Arbeiten, von denen hier nur jene von WATERHOUSE (115), LICHTENSTEIN (50), WEISKRANTZ (119) und von PRIBRAM und WEISKRANTZ (93) erwähnt seien, vermögen einen Einblick in die Komplexität dieser speziellen Fragestellung zu geben, die hier nicht Gegenstand weiterer Erörterung sein soll. Zusammenfassend kann also über die Versuche zu einer lokalisatorischen Aufteilung des Frontalhirnsyndroms beim Macacen gesagt werden, daß nur dem Defizit der aufgeschobenen Reaktion ein engeres Fokalfeld zugewiesen werden kann, nicht aber der Hyperaktivität oder der Störung der Emotionalität.

Strukturen außerhalb des Frontallappens und aufgeschobene Reaktion. Viele der schon besprochenen Arbeiten über die Wirkung der Entfernung von posteriorem Cortex auf das Erlernen von Sinnesassoziationen testeten die Tiere nach den Operationen auch auf die Fähigkeit zu aufgeschobenen Reaktionen. In keiner Arbeit wurde ein Defizit dieser Funktion nach der Entfernung von posteriorem Cortex nachgewiesen. Ein anderes Ergebnis erbrachten dagegen die Untersuchungen über die Bedeutung subcorticaler Strukturen für diese komplexe Leistung. Einen ersten Hinweis darauf, daß subcorticale Strukturen an der Leistung der aufgeschobenen Reaktion mitbeteiligt sind, ergab eine Studie von WADE (113), wo der Effekt der Circumcision des frontalen Cortex mit der Lobotomie verglichen wurde. Die Trennung des frontalen Cortex vom benachbarten Cortex durch die Circumcision beeinträchtigte die aufgeschobene Reaktion nicht, im Gegensatz zu der Lobotomie, bei der die Verbindungen zum Subcortex durchtrennt wurden. Da zwischen dem frontalen Cortex, insbesondere zwischen dem dorsolateralen Cortex entlang des Sulcus principalis und der parvocellulären Portion des Nucleus dorsomedialis des Thalamus ausgedehnte Projektionen vorhanden sind, lag es nahe, die Beteiligung dieses Kerngebietes zu der in Frage stehenden Funktion abzuklären. CHOW (16), sowie PETERS et al. (84) setzten in dieser Struktur ausgedehnte stereotaxische elektrolytische Ausschaltungen, doch trat in keinem Fall der Verlust der aufgeschobenen Reaktion ein. Nach den Arbeiten von HARMAN et al. (35) und von METTLER et al. (55) bestehen vom frontalen Cortex im weiteren auch direkte anatomische Verbindungen zu dem Nucleus caudatus. Daß der Nucleus caudatus in der Tat für die Funktion der aufgeschobenen Reaktion von Bedeutung ist, zeigen die Arbeiten von ROSVOLD und DELGADO (98) und von ROSVOLD et al. (99), die beide nachwiesen, daß nach Koagulation innerhalb des Kopfes des Nucleus caudatus die Tiere im Test der aufgeschobenen Alternation vermehrt Fehler begehen. Aus den weiteren Arbeiten von BÄTTIG und ROSVOLD (3) und von MIGLER (58) ergibt sich ferner die Wahrscheinlichkeit, daß das Defizit im aufgeschobenen Reagieren um so größer wird, je größere Ausschaltungen im Nucleus caudatus gesetzt werden, und daß bei vollständiger Zerstörung des Kernes ein Defizit zu erwarten ist, das jenem, an Totalität nicht nachsteht, welches nach Ausschaltung des dorsolateralen frontalen Cortex auftritt. Diese Erwartung wurde übrigens neuerdings durch Dean und Davis (19) voll bestätigt. Nach subtotaler Zerstörung des Caudatus fanden diese Autoren einen völligen Verlust der Fähigkeit zur aufgeschobenen Reaktion beim Macacen. Aus der Arbeit von Migler (58) wird es auch sehr wahrscheinlich, daß Durchtrennung des vorderen Teiles des Corpus callosum die Fähigkeit zur aufgeschobenen Reaktion ebenfalls beeinträchtigt. Da dieser Autor den Nucleus caudatus von medial her nach Durchtrennung des Corpus callosum durch Absaugen entfernte, testete er in seiner Studie auch operierte Kontrolltiere mit einer einfachen Durchtrennung des Corpus callosum. Diese operierten Kontrolltiere erlernten in der Folge die aufgeschobene Reaktion wohl schneller als caudatektomierte, jedoch langsamer als nicht-operierte Kontrolltiere. Diese Tatsache erhält erhöhte Bedeutung durch den Befund Rosvolds (101), daß durch das Corpus callosum gekreuzte Fasern vom dorsolateralen frontalen Cortex nach dem gegenseitigen Nucleus caudatus verlaufen. Zusammenfassend kann geschlossen werden, daß außerhalb des dorsolateralen frontalen Cortex kein anderer Cortexteil des Macacenhirns an der Funktion der aufgeschobenen Reaktion beteiligt ist, daß aber der Nucleus caudatus für diese Funktion eine wichtige Rolle spielt und daß damit auch die Frage gestellt werden muß, inwiefern weitere subcorticale Strukturen an der Funktion mitbeteiligt sein könnten.

Psychologische Ursachen des Versagens frontalhirnoperierter Affen in der aufgeschobenen Reaktion. Eine Reihe von Einzelfaktoren, die für dieses Versagen operierter Affen verantwortlich sein könnten, war Gegenstand einer Vielzahl von Untersuchungen. Die Tatsache, daß die Zeit von der Reizpräsentation bis zur Reaktion nicht überbrückt werden kann, ließe daran denken, daß die operierten Tiere einen Gedächtnisverlust haben. Doch wurde schon dargelegt, wie das Erlernen von visuellen Unterscheidungen, bei denen ein Tier immerhin von einem Tag zum anderen das Erlebte nicht vergessen darf, dem frontalhirngeschädigten Affen keine größere Schwierigkeit bietet als dem normalen Tier. Das Verhaltensdefizit der frontalhirngeschädigten Tiere erhielt daher Bezeichnungen wie Verlust von „Kurzgedächtnis" oder vom „Sofortgedächtnis", womit angedeutet wird, daß neben dem Gedächtnis im üblichen Sinn eine zweite Form von Gedächtnis zur Überbrückung sehr kurzer Zeitspannen postuliert wird. Von dieser Hypothese aus wäre zu erwarten, daß eine aufgeschobene Reaktionsaufgabe einem operierten Affen um so mehr Mühe bereiten würde, je länger der Aufschub ist. Meyer et al. (56) verglichen die Leistungsfähigkeit bei Aufschubzeiten von 5, 10, 20 und 40 sec und fanden, daß operierte Affen versagten, egal wie lange die Aufschubzeit war, und egal ob versucht wurde, die Aufschubzeit sukzessive von 5 auf 40 sec zu steigern, oder ob kurze und lange Aufschubszeiten einander in zufälliger Reihenfolge ablösten. Aus einer Studie von MISHKIN und WEISKRANTZ (71) läßt sich dagegen schließen, daß diese Folgerung für Aufschubszeiten unter 5 sec nicht gilt. Die operierten Tiere waren in dieser Untersuchung fähig, Aufschubszeiten zu überbrücken, wenn sie langsam von 0 auf 8 sec gesteigert wurden, nicht aber, wenn abrupt ein Aufschub von 8 sec eingeführt wurde. Es muß aber gesagt werden, daß in dieser Untersuchung zwischen Reiz und Reaktion kein Schirm vor das Tier gesenkt wurde, und daß nicht die Reaktion selber, sondern die der Reaktion folgende Belohnung aufgeschoben wurde. Der Befund erweitert also die alte Kenntnis vom Versagen in aufgeschobenen Reaktionen. Auch ein Aufschub der Belohnung statt der Reaktion läßt frontalhirnoperierte Affen versagen. Auf die Frage inwieweit es kritisch ist, daß vor dem Tier während dem Aufschub ein Schirm heruntergesenkt wird, um ihm die Sicht auf Reiz und Reaktionsort zu unterbrechen, findet sich eine Antwort in einer Arbeit von BÄTTIG et al. (3). Hier wurde zuerst immer während des Aufschubs ein Schirm vor dem Tier gesenkt und der Aufschub selber graduell von 0 auf 5 sec gesteigert. Sobald der Schirm gesenkt wurde, und wenn es auch nur ein momentanes Hinunterlassen und Hochziehen, ein Aufschub von „0 sec" war, versagten die operierten Affen vollkommen, auch wenn das Training mit jedem Tier über Wochen hinweg fortgeführt wurde. Wurde aber der Schirm nicht gesenkt, und die Tiere während dem Aufschub an der Reaktion nur dadurch gehindert, daß der Experimentator die Kartondeckel über den Futterlöchern mit seinen Händen festhielt, so meisterten die operierten Tiere alle Aufschubzeiten bis zu 5 sec. Jedoch brauchten sie auch zu dieser erleichterten Form der aufgeschobenen Reaktion viel mehr Versuche als von einem normalen Affen erwartet würde. Die Erklärung dafür ergab sich aus der Beobachtung der Versuchstiere. Diese waren nur dann fähig, nach dem Aufschub korrekt zu reagieren, wenn sie „sich zwangen", während des Aufschubs hinter dem korrekten Futterloch sitzen zu bleiben. Dieses Sitzenbleiben hatten sie vorerst zu erlernen, was ihnen um so mehr Mühe bereitete, da sie alle hyperaktiv waren. Für Normaltiere bereitet dieses Problem keine Mühe. Sie sind erstens nicht hyperaktiv und zweitens fähig, auch dann den korrekten Futterplatz ohne weiteres wieder zu finden, wenn sie während des Aufschubs im Käfig herumgeturnt haben. Damit läßt sich schließen, daß es operierten Macacen unmöglich ist, einen echten Aufschub, bei dem sie physisch vom Belohnungs- und Reaktionsort getrennt sind, zu überbrücken. Sind sie nicht physisch vom Ort der Reaktion und vom Reiz getrennt, so erlernen sie sich eine Art „Eselsbrücke" zu verschaffen, indem sie vor dem korrekten Reiz sitzen bleiben, bis der Aufschub vorbei ist, womit auch nicht mehr von einem echten Aufschub gesprochen werden kann, weil der vor dem Aufschub gebotene Reiz durch das Sitzenbleiben am korrekten Ort substituiert wird. Somit bleibt das schon erwähnte Konzept des „Kurz"- oder „Sofortgedächtnisses", so unbefriedigend es auch in seiner Formulierung anmuten mag, weiterhin in Gültigkeit. Als weiterer Faktor wurde untersucht, inwieweit eventuell die ruhelose Aktivität frontalhirngeschädigter Affen einen Anteil am Versagen in der aufgeschobenen Reaktion haben könnte. Von diesem Gesichtspunkt aus betrachtet würde dann nicht mehr von einem Verlust von Kurzgedächtnis gesprochen, sondern von einem Verlust an Aufmerksamkeit und „Konzentrationsfähigkeit", bedingt durch das dauernde Reagieren auf jeden geringsten interkurrenten Reiz. MEYER et al. (56) glaubten einen solchen Zusammenhang ausschließen zu dürfen, da sie beobachteten, daß die einseitige Entfernung des frontalen Cortex eine deutliche Hyperaktivität, nicht aber ein Versagen in der aufgeschobenen Reaktion bewirkte. Auch HARLOW et al. (34) kamen zum analogen Schluß und stellten zudem fest, daß das spezifische Versagen der operierten Affen nicht nur zur Hyperaktivität, sondern auch zu operationsbedingten Veränderungen in der Emotionalität keine Beziehungen aufweise. Was die Rolle der Hyperaktivität im Verhalten der hirngeschädigten Affen sei, kamen jedoch spätere Autoren zu anderen Schlüssen. MALMO (52) setzte das Ausmaß der Aktivität während des Aufschubs in Beziehung zu Erfolg oder Mißerfolg der nachfolgenden aufgeschobenen Reaktion. Er fand, daß eine falsche Reaktion des Versuchstieres um so eher auftrat, je aktiver das Tier in der der Reaktion vorangehenden Aufschubszeit war. Er konstatierte ebenfalls, daß die Tiere während dem Aufschub aktiver waren, wenn sie den Aufschub in einem erleuchteten, statt in einem verdunkelten Käfig zubrachten. WADE (112), PRIBRAM (86) und MISHKIN et al. (63) verfolgten diese Idee konsequenterweise weiter, indem sie versuchten, die Leistungsfähigkeit operierter Tiere durch die Applikation von sedativ wirkenden Barbituraten zu steigern. Alle drei Untersuchungen fanden eine eindeutig verbesserte Leistungsfähigkeit der operierten Affen unter der Barbituratwirkung, wenn auch in keinem Fall die Leistung nicht operierter Tiere nur annähernd erreicht wurde. DEAN und DAVIS (19) und DAVIS (18) fanden das gleiche Ergebnis nicht nur an frontalhirngeschädigten, sondern auch an Nucleus caudatus-geschädigten Affen. Interessanterweise fanden sie aber diese Wirkung nicht nur als Folge der Applikation von sedativen Barbituraten, sondern auch als Folge der Gabe exzitierender Stoffe wie Phenyldiäthyldiacetat. Diese Zusammenhänge nahmen STANLEY und JAYNES (109) in einer Übersichtsarbeit zusammen mit anderen Befunden zum Anlaß, das Verhaltensdefizit frontalhirnoperierter Affen als einen Verlust von Hemmfähigkeit aufzufassen. Operierte Affen wären demnach nicht mehr fähig, Reaktion und Aufmerksamkeit auf interkurrierende afferente Reizung während des Aufschubs zu hemmen, wodurch die „Erinnerungsspur", die für eine korrekte aufgeschobene Reaktion notwendig ist, ausgewischt würde. Diese Autoren stellten auch die Hypothese auf, daß dieser Verlust an

Hemmfähigkeit auf die Beschädigung der frontalen „Suppressor Areale", wie sie zu jener Zeit auf Grund elektrophysiologischer Studien postuliert wurden, zurückzuführen sei.

Ein weiterer Punkt, der untersucht wurde, ist die Art der Motivation zur Reaktion. Während alle bisher erwähnten Autoren ihre Tiere für eine korrekte Reaktion mit Futter belohnten, belohnten MILES und ROSVOLD (59) ihre Tiere dadurch, daß sie ihnen für eine korrekte Reaktion keine elektrischen Schläge gaben, während sie falsche Reaktionen mit solchen Schmerzreizen bestraften. Da die operierten Tiere auch unter diesen Bedingungen versagten, kann nicht angenommen werden, daß die Operation eine Abschwächung der Futtermotivation und durch diese sekundär ein Versagen in aufgeschobenen Reaktionen bewirke.

Die Hypothese, daß frontalhirnlädierte Affen an Unterscheidungs- und Erkennungsfähigkeit für den dem Aufschub vorausgehenden Reiz einbüßen, untersuchten weitere Arbeiten im Test der aufgeschobenen Reaktion. MISHKIN und PRIBRAM (68) fanden, daß es keinen Einfluß auf die Leistung der operierten Tiere hatte, ob den Tieren die Belohnung vor dem Aufschub direkt an einem bestimmten Platz gezeigt worden war (direkte Methode), oder ob die korrekte Stelle nur mit einem „Futtersignal" (irgend ein Gegenstand) dem Tier vor dem Aufschub angezeigt wurde (indirekte Methode). Dagegen konnten die Autoren die Aufgabe für die operierten Tiere erleichtern, wenn sie pro Wahlsituation nur einen, statt zwei Reizen verwendeten. Diesen einen Reizgegenstand stellten sie auf ein einziges vorhandenes Wahlloch in der Mitte, oder wenn zwei Wahllöcher vorhanden waren, in die Mitte zwischen die beiden. Im ersten Fall sagte die Qualität des Reizes, ob in dem einzigen Wahlloch Belohnung sei oder nicht, im anderen Fall, ob die Belohnung im Wahlloch links oder rechts vom Reiz liege. Von diesen beiden Aufgaben bedeutete nur jene eine Erleichterung, bei der nur ein einziges Wahlloch vorhanden war. Daraus müßte man schließen, daß frontal operierte Affen besondere Mühe haben, die Bedeutung von „links und rechts" über einen zeitlichen Aufschub hinweg „im Kopfe zu behalten". Zu einem analogen Schluß kamen schon MEYER et al. (56), die ebenfalls folgerten, daß frontal operierte Tiere die positionelle Bedeutung eines Reizes nicht über die Zeit des Aufschubes hinüberretten könnten. Diese Erklärung wäre an und für sich sehr einleuchtend, wenn man sich das Verhalten solcher Affen vor Augen hält, die infolge ihres Zwangslaufens im Kreise herum sicher Schwierigkeiten haben, die räumliche Bedeutung von links und rechts zu verwerten. Doch kann der Schluß aus den beiden erwähnten Arbeiten nicht mit Sicherheit gezogen werden, da es immer noch möglich wäre, daß andere Faktoren eine kritische Rolle mitspielen würden, wie z.B. die Tatsache, daß ein einziger in der Mitte des Wahlbrettes gelegener Reiz eine einfachere Reizsituation darstellt, als zwei Reize, ein negativer und ein positiver Reiz links und rechts auf dem Wahlbrett. PRIBRAM und MISHKIN

(92) konnten die Möglichkeit, daß die Operation den Tieren die Fähigkeit nehme, die Links-Rechts-Bedeutung von Reizen über den Aufschub hinweg „im Kopfe zu behalten" jedenfalls eindeutig mit einer einleuchtenden Testanordnung ausschalten. Die Autoren legten den Tieren zwei Gegenstände verschiedener Art zufällig vertauscht auf das linke und rechte Futterloch. Egal wo die Gegenstände lagen, hatten die Tiere abwechslungsweise den Gegenstand A und B zu wählen. Links oder rechts hatte hier keine Bedeutung mehr, und die Aufgabe bestand nur darin, jenen Gegenstand zu vermeiden, unter dem im vorangegangenen Versuch Futter lag. Die operierten Tiere versagten auch bei dieser Anordnung vollkommen, und ihre Leistung verbesserte sich im Verlaufe langen Trainings kaum merklich. Aus diesen Arbeiten muß der Schluß gezogen werden, daß für das Versagen die Erkennbarkeit des vor dem Aufschub gebotenen Reizes wohl eine wichtige Rolle spielt, daß aber der entscheidende Faktor für diese Erkennbarkeit noch im Dunkel liegt.

Einige weitere Untersuchungen versuchten im Test der aufgeschobenen Reaktion die Reaktion an sich zu variieren. Mishkin und Pribram (67) ließen die operierten Tiere statt nach links oder rechts, nach oben oder nach unten greifen und in einer dritten Testart verlangten sie von ihnen alternierend einmal nach einem einzigen Futterplatz zu greifen, und im nächsten Versuch nicht. Die operierten Affen erlernten einzig das alternative Greifen und Nichtgreifen, doch läßt sich dieses Ergebnis wie schon gesagt auch dadurch erklären, daß bei dieser Testart ja nicht nur die Reaktion an sich, sondern auch der vorausgehende Reiz in kritischer Weise verändert worden war. Somit bleibt die Frage offen, ob auch die Art der Reaktionen neben den Eigenschaften des vor dem Aufschub gegebenen Reizes kritische Faktoren für das Versagen operierter Affen enthält.

Als weitere Möglichkeit wurde vorgeschlagen, daß operierte Affen eine geringere Futtermotivation besäßen, oder daß sie durch unbelohnte Fehlwahlen rascher entmutigt würden. Finan (24) stellte diese Hypothese auf, nachdem er fand, daß operierte Tiere ihre Leistung verbesserten, wenn er sie vor dem Aufschub am Orte wo nach dem Aufschub die korrekte Reaktion zu erfolgen hatte, vorfütterte. Jedoch kann diese Wirkung auch darauf zurückgeführt werden, daß bei diesem Vorgehen der vor dem Aufschub gebotene Reiz verstärkt wurde, indem die Affen vor dem Aufschub nicht nur sahen, daß die Belohnung z.B. links hingelegt wurde, sondern auch auf diese Stelle durch die Vorausbelohnung stärker hingelenkt wurden. Daß diese Erklärung angenommen werden muß, belegt die Arbeit von Harlow et al. (34), die den Tieren im Verlaufe des Testes zusätzliche Belohnungen gaben, die in keinem Zusammenhang zum Test selber standen, wobei sich die Leistung der operierten Affen nicht verbesserte. Zusammenfassend läßt sich aus diesen Arbeiten schließen, daß das Versagen frontalhirngeschädigter Affen bei der aufgeschobenen Reaktion in der Unfähigkeit beruht, kurze Zeitspannen zu überbrücken. Die als Folge der Hyperaktivität herabgesetzte Aufmerksamkeit und der Gehalt des dem Reiz vorangehenden Reizes stellen zusätzliche Faktoren dar. Eine genauere psychologische Definition dieses Versagens läßt sich bis heute nicht geben.

Zeitliche Dauer des Verhaltensdefizites frontalhirnoperierter Affen. Einheitlich kann aus einer Reihe von Arbeiten geschlossen werden, daß es nach totaler Entfernung des frontalen Cortex bei erwachsenen Tieren auch nach sehr langer Zeit zu keiner Remission der Fähigkeit, aufgeschobene Aufgaben zu lösen, kommt. In der bereits erwähnten Arbeit von HARLOW et al. (34) zeigte sich 8 Monate nach der Operation noch keine solche Remission. BRUSH und MISHKIN (9) untersuchten gar Tiere, bei denen die Operation Jahre zurücklag, mit dem gleichen Ergebnis. Eine andere Frage stellt sich danach, wie früh das Verhaltensdefizit nach der Operation zum Vorschein komme. FORGAYS (25) hatte auf Grund eigener Experimente an Ratten und auf Grund klinischer Literatur postuliert, daß ein Effekt nach Beschädigung von Hirnsubstanz erst mit einer gewissen Latenz auftrete. ORBACH (79) untersuchte diese Hypothese näher, indem er Affen zweizeitig operierte. In der ersten Sitzung erhielten die Tiere nur die Bohrlöcher im Schädel und in der zweiten Sitzung wurde durch diese Bohrlöcher in einem schwachen Ätherrausch die frontale Lobotomie ausgeführt. Das Vorgehen erlaubte, die Tiere schon $1^{1}/_{2}$ Std nach dem entscheidenden Eingriff zu testen, ohne auf die Erholung von einer tiefen Operationsnarkose warten zu müssen. Der Autor fand den Volleffekt der Operation schon nach diesen $1^{1}/_{2}$ Std nach dem Eingriff, und schließt daher, daß die Ergebnisse von FORGAYS und anderen Autoren durch den Umstand entstanden, daß ihre Tiere nach der Operation noch längere Zeit unter der Barbituratwirkung der Narkose standen. Diese Annahme erscheint besonders glaubhaft, weil Barbiturate die Leistungsfähigkeit operierter Affen verbessern können, wie schon weiter oben besprochen wurde.

Die sensorische Unterscheidungsfähigkeit frontalhirnoperierter Affen. JACOBSEN (37) und schon lange vor ihm BIANCHI (5) und FERRIER (23) nahmen an, daß der anteriore frontale Cortex ausschließlich komplexen psychischen, jedoch nicht assoziativen sensorischen Funktionen diene, während der posteriore Cortex als der eigentliche assoziative sensorische Cortex des Affen anzusehen sei. Diese Ansicht blieb über lange Zeit hindurch auch unbestritten, da ihr keine experimentellen Ergebnisse widersprachen oder zu widersprechen schienen. Nach Ausschaltung des frontalen Cortex wurde immer wieder die Unfähigkeit beobachtet, Reaktionen aufzuschieben und nach Ausschaltung des posterioren Cortex immer Störungen des sensorischen Unterscheidungslernens.

In neuerer Zeit konnte jedoch vermehrt gezeigt werden, daß unter speziellen Bedingungen auch Störungen des Unterscheidungslernens bei frontalhirnoperierten Affen auftreten, und daß ähnliche Befunde schon aus früheren Arbeiten hervorgehen, ohne daß ihnen die genügende Beachtung zukam. Da aber heute mehr Klarheit über die Natur solcher Befunde besteht, sei vorweggenommen, daß diese Verhaltensdefizite nur scheinbar sensorischer Natur sind. Die fraglichen Verhaltensdefizite müssen auf einem Ausfall anderer und wahrscheinlich höherer komplexerer Funktionen beruhen, ähnlich wie die Unfähigkeit, Reaktionen aufzuschieben. In einer relativ frühen Arbeit von Pribram et al. (87) fällt auf, daß frontalhirnoperierte Affen visuelle Unterscheidungen nach der Operation langsamer erlernten, als normale Tiere, ohne daß diesem Befund weiter Beachtung geschenkt worden wäre. Harlow und Dagnon (32) fanden ähnlicherweise, daß Affen nach der Operation mehr Mühe hatten, von der einen auf eine andere und zwar diametral entgegengesetzte Bedeutung von visuellen Signalen umzulernen als unoperierte Affen. Diese Autoren sahen darin eher einen Verlust einer höheren intellektuellen Funktion als ein sensorisches Unterscheidungsdefizit.

In einer neueren Arbeit von Bättig et al. (4) findet sich ein direkterer Hinweis auf die Natur solcher visueller Unterscheidungsdefizite frontalhirnoperierter Affen. Hier erlernten operierte Affen eine gewöhnliche visuelle Unterscheidung normal in der einen methodischen Testvariante und verlangsamt in der anderen Testvariante. Die klassische simultane Unterscheidung von zwei gleichzeitig auf dem Wahlbrett liegenden visuellen Mustern bereitete den Tieren keine Mühe. Wurden aber die Reize sukzessiv präsentiert, indem in einzelnen Versuchen der positive Reiz allein dalag und das Futter bedeckte, während in anderen Versuchen die negative Karte auf dem leeren Futterloch lag, hatten die gleichen Tiere große Mühe, sich beim Anblick der negativen Karte davon zu enthalten, das Futterloch trotzdem abzudecken. Interessanterweise findet sich in dieser Arbeit das gleiche Syndrom auch für Nucleus caudatus operierte Tiere in schwächerer Form, womit eine Parallele zu den Untersuchungen über die Fähigkeit zur aufgeschobenen Reaktion nach Ausschaltung des frontalen Cortex und des Nucleus caudatus besteht. Ganz ähnlich konnten auch Brush und Mishkin (9) zeigen, daß ein Versagen frontalhirnoperierter Affen in Unterscheidungstesten als Folge abgewandelter methodischer Vorgehen auftreten kann. Diese Autoren trainierten die Affen, in einem monatelangen Serientest eine große Zahl von verschiedenen Gegenständen zu unterscheiden. Täglich kamen neue Gegenstände an die Reihe und in jeder Sitzung wurden nur wenig Einzelversuche gegeben. Im ersten Einzelversuch lagen immer zwei Gegenstände auf dem Wahlbrett, wobei entweder unter beiden Gegenständen oder unter keinem von beiden Futter lag. Unter dem in diesem ersten Einzelversuch spontan gewählten, vorher noch nie gesehenen Gegenstand lag nun in allen folgenden Einzelversuchen ebenfalls Futter, wenn das im ersten Einzelversuch der Fall war, oder es lag umgekehrt nie Futter darunter, wenn im ersten Versuch ebenfalls kein Futter darunter gelegen hatte. Die Variante, daß unter dem initial gewählten Gegenstand immer Futter ist, konnten die frontalhirnoperierten Affen ebenso schnell lernen, wie die unoperierten Affen. Dagegen erlernten sie die andere Variante, daß unter diesem Gegenstand nichts zu finden sei, viel langsamer. Der erste Befund von BÄTTIG ließe sich dahin interpretieren, daß frontalhirnoperierte Tiere ein Defizit an Hemmung in der Terminologie von PAVLOV haben, da sie nur langsam lernen, die Reaktion auf e inen unbelohnten negativen Reiz zu hemmen. Die Studie von BRUSH und MSHKIN könnte ähnlich interpretiert werden, indem operierte Tiere eine erhöhte Perseveration für einen initial bevorzugten Reiz hätten. In beiden Fällen läßt sich ein echtes sensorisches Unterscheidungsdefizit ausschließen, da ja die Affen bei entsprechenden methodischen Voraussetzungen imstande waren, dieselben Gegenstände und bemalten Karten zu unterscheiden.

Ähnliche Parallelen ergeben sich auch für andere Sinnesmodalitäten. WEISKRANTZ und MISHKIN (120) fanden, daß operierte Affen korrektes Reagieren auf sukzessiv präsentierte auditorische Reize langsamer erlernten als unoperierte Affen. In der schon erwähnten Arbeit von BÄTTIG et al. (4) wurde mit derselben Technik ebenfalls der Verlust der auditorischen Unterscheidungsfähigkeit nachgewiesen. Jedoch lernten die gleichen Affen die gleichen Töne zu unterscheiden, wenn sie in einem anderen auditorischen Test lernen mußten, für das eine Signal einen links angeordneten und für das andere Signal einen rechts angeordneten Hebel zu pressen (automatischer WGTA). BLUM (7) fand auch mit diesem Vorgehen der simultanen Links- oder Rechtswahl ein Lerndefizit der operierten Tiere, doch verwendete er als zusätzliche Motivation zum Futter elektrische Schläge, mit denen er unkorrekte Ortswahlen bestrafte. Dieses Ergebnis ließe sich im gleichen Sinne wie die schon besprochenen erklären, indem der Schock bei den normalen Tieren, die ein normales Hemmvermögen haben, rascher die unkorrekten Reaktionen unterdrückte, als bei den geschädigten operierten Tieren. Diese Interpretation gewinnt sehr an Wahrscheinlichkeit, wenn man eine Studie von WATERHOUSE (115) zum Vergleich heranzieht. Dieser Autor trainierte die Tiere, auf visuelle oder akustische Signale von zwei Futterbehältern entweder den links oder rechts liegenden zu öffnen. Öffnen des korrekten Behälters wurde belohnt, Öffnen des falschen Behälters wie in BLUMs Untersuchung mit elektrischen Schlägen bestraft. Unter dieser Voraussetzung machten die operierten Tiere nicht nur in der Beantwortung akustischer, sondern auch in der Beantwortung visueller Signale mehr Fehler als die Kontrolltiere. Auch bei der Unterscheidung von Tastreizen lassen sich ähnliche komplexe Verhältnisse beobachten. Der Test wird gewöhnlich so ausgeführt, daß die Tiere eine Futterkiste nur berühren, nicht aber sehen können, und auf Grund der Oberflächenqualität des Kistendeckels entscheiden müssen, ob sie den Behälter öffnen wollen oder nicht. Da die Tiere, auch wenn zwei Behälter daliegen, gewöhnlich zur gleichen Zeit nur einen der beiden Behälterdeckel mit beiden Händen betasten, liegen die Verhältnisse ähnlich wie bei der sukzessiven Testart für die Unterscheidung von Sinnesreizen anderer Modalitäten. Ettlinger und Wegener (21) fanden, daß frontal operierte Tiere eine solche Aufgabe langsamer erlernten als unoperierte Affen. Rosvold et al. (100) variierten die Methode so, daß die Kistendeckel viel schwerer zu öffnen waren, und dabei lernten die operierten Affen wieder ebensoschnell, wie unoperierte Affen. Allen (1) fand ferner auch, daß Hunde mit präfrontalen Ausschaltungen weniger feine Geruchsunterscheidungen treffen konnten. Auf Grund der bisherigen Besprechung könnte man erwarten, daß auch dieses Ergebnis in ähnlicher Weise unspezifisch sei, obwohl der experimentelle Nachweis dazu bis heute fehlt. Brutkowski, Konorski et al. (10) studierten die Natur solcher Verhaltensdefizite systematisch beim Hunde. Sie trainierten die Tiere vor der Operation, Reaktionen auf gewisse Tonreize zu unterlassen, und auf andere Tonreize hin auszuführen. Nach der Operation machten die Tiere viele Fehler, aber nur indem sie auf die unbelohnten negativen Reize hin wieder reagierten. Die Autoren sehen in diesem Defizit ebenfalls einen Verlust des Hemmvermögens. Dieser Verlust war aber, im Gegensatz zu dem Verlust, Reaktionen aufzuschieben nur temporär und weniger krass. Die Autoren glauben daher, der Verlust, Reaktionen aufzuschieben und der Verlust an Hemmvermögen sei auf die Schädigung von zwei voneinander unabhängigen Funktionen zurückzuführen, und eventuell sogar innerhalb des frontalen Cortex an lokal voneinander getrennte Fokalgebiete gebunden. Weitere Autoren der gleichen polnischen Forschergruppe stützen diese Annahmen mit verschiedenem methodischen Vorgehen. Lawickas (49) Hunde hatten zu bellen, um Futter zu bekommen, doch nützte das Bellen nur, wenn es nach dem positiven Reiz ausgeführt wurde. Nach der Operation bellten die Hunde auch wieder nach unbelohnten negativen Reizen. Brutkowski (11) erhielt analoge Ergebnisse mit klassischen bedingten Speichelreflexen nach der Technik Pavlovs.

Aus all diesen Arbeiten geht hervor, daß zwar die Ausschaltung des frontalen Cortex die Unterscheidungsfähigkeit für sensorische Reize nicht schädigt, daß aber umgekehrt diese Operation nicht nur das Vermögen, Reaktionen aufzuschieben, löscht. Das Frontalhirnsyndrom des Affen und des Hundes ist vielmehr bedeutend komplexerer Natur als ursprünglich angenommen wurde. Wenn auch für eine Reihe von Befunden ein Verlust an Hemmungsvermögen als Erklärung angenommen werden könnte, so läßt sich doch diese Erklärung für andere Fälle nur mit Mühe aufrecht erhalten, wie einige weitere Beispiele zeigen.

Mishkin und Weiskrantz (72) bestimmten bei operierten Affen die kritische Frequenz bei der einzelne kurze Lichtblitze subjektiv zu einer kontinuierlichen Lichtquelle verschmelzen. Die Autoren fanden, daß diese kritische Verschmelzungsfrequenz bei operierten Affen höher liegt als bei Kontrolltieren. Für diesen Fall ließe sich die Theorie eines Hemmverlustes wohl nur mit Hilfe spekulativer Erklärungsversuche aufrecht erhalten. MEYER und SETTLAGE (57) beobachteten die Tiere in ihrem Verhalten, wenn sie vier Futterkisten vor sich hatten, von denen vollkommen zufällig bald die eine, bald eine andere, Futter enthielt. Beim Suchen war das Verhalten der operierten Tiere weniger stereotyp als das der Kontrolltiere. So versuchten die operierten Tiere mit viel weniger Ausdauer als die Kontrolltiere eine bestimmte Kiste zu öffnen, wenn dieses systematische Spiel nicht zum Ziele führte. Trotzdem die operierten Tiere weniger stereotyp an solchen Hypothesen zur Lösung des unlösbaren Problems festhielten, waren sie fähig, eine schon probierte, aber erfolglose Hypothese nicht zu wiederholen. FRENCH (27) fand, daß operierte Affen sich weniger an einem Hebel zu schaffen machen, der in ihrem Käfig angebracht ist, und dessen Betätigung entweder gar keine Folgen hat, oder die Käfigbeleuchtung verstärkt. Auch diese beiden Resultate ließen sich nur mühevoll nach dem Konzept des Hemmverlustes erklären. Eine Brücke zu der Hyperaktivität oder dem Verlust der Fähigkeit, Reaktionen aufzuschieben, zu schlagen, dürfte vom psychologisch theoretischen Standpunkt aus ebenfalls sehr schwierig sein.

VIII. Diskussion und Hinweis auf die Befunde am hirngeschädigten Menschen

Als wichtigstes Faktum läßt sich aus allen diesen Arbeiten schließen, daß der höhere Cortex des Affen wenigstens bis zu einem gewissen Grade tatsächlich in funktionsspezifischer Weise spezialisiert ist. Wie schon JACOBSEN (37) postulierte, müssen dem anterioren Cortex höhere komplexe psychische Funktionen zugeschrieben werden, während dem posterioren Cortex assoziative sensorische Funktionen zukommen. In beiden Fällen ist aber eine genaue Unterteilung des Cortex in Fokalgebiete einzelner psychischer Funktionen nicht überall möglich. Nur für die somatosensorische und visuelle Unterscheidungsfähigkeit können mit relativer Sicherheit circumscripte corticale Felder angegeben werden. Das Fokalgebiet für Geruchs- und Geschmacksunterscheidungen überschneidet sich eventuell mit dem Cortexareal für visuelle Unterscheidungen. Auf ein spezifisches Hirnfeld für auditorische Unterscheidungen beim Affen läßt sich bis heute nicht schließen. In keinem der Fälle findet ferner der Mechanismus der Erholung einer assoziativen Funktion nach Beschädigung des kritischen Hirnfeldes eine Abklärung. Das psychische Frontalhirnsyndrom des Affen ist wesentlich komplexer als ursprünglich angenommen wurde. Der psychologische Hintergrund des auffälligsten Defizites frontalhirnoperierter Affen, nämlich das irreparable Versagen in aufgeschobenen Reaktionstesten, ist bis heute noch nicht abgeklärt. Die komplexe und vielfältige Natur des Frontalhirnsyndroms des Affen zeigt deutlich, wie entweder eine Reihe von Funktionen im selben Hirnfeld vertreten sein können, oder wie wenig eine eventuelle funktionelle Zusammengehörigkeit solcher verschiedener Leistungen bis heute psychologisch abgeklärt werden konnte. Diese Tatsachen sprechen zum Teil gegen die starre anatomisch funktionelle Konzeption, die hinter vielen der besprochenen Arbeiten steht.

Neben den neuen Erkenntnissen, die in vielen Fällen auf Grund früherer anatomischer und physiologischer Studien aufgestellte Hypothesen erstmals experimentell wissenschaftlich belegen konnten, steht als großer Fortschritt die sukzessive Verfeinerung der psychoexperimentellen Methodik im Tierversuch. Diese Methoden erlauben es, die verschiedensten psychischen Funktionen in gezielter Weise zu testen. Damit ist ein Rüstzeug im Werden, das sicher der Psychophysiologie und der immer mehr in den praktischen Vordergrund drängenden Psychopharmakologie zu Fortschritten verhelfen kann, die früher nicht möglich gewesen wären. Die Fortschritte dieser Weiterentwicklung der psychoexperimentellen Methodik liegen in einer Reihe von Einzelfaktoren. Dem Experiment am gefesselten Tier, wie es Pavlov und seine Schüler ausführten, ist das Experiment am frei beweglichen Tier gefolgt. Die starre Methodik der Zeit Pavlovs wurde aufgelockert und das einzelne Experiment ist mehr und mehr direkt auf das Ziel der untersuchten psychischen Funktion gerichtet und weniger auf abstrakte psychologische Funktionstheorien. Das Tier wird nicht mehr als ein willenloser Apparat aufgefaßt, dessen Verhalten bestimmt ist durch passiv aufgenommene Sinneseindrücke und deren gesetzmäßige Verarbeitung im Zentralnervensystem, sondern als ein Organismus, dem eine endogene „Willens- und Entschlußkraft" und in begrenzter Weise sogar eigene „Überlegungskraft" zugestanden wird.

Die Besprechung dieser Arbeiten wäre unvollständig ohne einen kurzen Hinweis auf die Ergebnisse von psychologischen Untersuchungen am hirnoperierten Menschen.

Leider ist ein solcher Vergleich als Folge mehrerer Umstände nur sehr schwer durchzuführen, und kann nicht die erwarteten Parallelen erbringen. Eine erste Schwierigkeit für Untersuchungen am Menschen liegt darin, daß die strengen Kontrollbedingungen des Experimentes am Affen in den wenigsten Fällen erfüllt werden können. Selten kann ein Mensch vor und nach einer Ausschaltung des Cortex mit der gleichen Methodik getestet werden. Ist die Schädigung des Cortex eine Unfallfolge, bestehen keine vorhergehenden Untersuchungen, ist sie eine chirurgische Behandlung, so war schon der funktionelle Zustand vor der Operation pathologisch. Selten trifft es sich, daß ganze Gruppen von Individuen mit der gleichen Hirntraumaanamnese verglichen werden können, und die Befunde an Einzelindividuen geben wenig statistisch haltbare Ergebnisse. Eine ebensogroße Schwierigkeit besteht darin, daß der Bildungsgrad, der auf die Leistungsfähigkeit in solchen Testen einen großen Einfluß hat, von Individuum zu Individuum sehr variabel ist. Neben diesen Schwierigkeiten, die in der Natur des Problems liegen, bestehen methodische Unterschiede in der bisherigen experimentellen Fragestellung beim hirngeschädigten Mensch und Tier. In den meisten Fällen untersuchte man den hirngeschädigten Menschen mit viel komplexeren Testen als dies beim Affen geschah. Die viel angewandten Intelligenzteste erfassen nicht in gezielter Weise eine einzelne psychologische Funktion, sondern Assoziationen aller verschiedenen Sinnesmodalitäten, Gedächtnis, höhere psychische Funktionen usw. Auf Grund dieser Tatsachen ist es nicht erstaunlich, daß wir über die funktionelle Organisation des menschlichen Cortex bis heute eigentlich weniger wissen als beim Affen. Daher seien in diesem Zusammenhang nur einige wenige experimentelle Arbeiten erwähnt, die methodisch den besprochenen Tierversuchen vergleichbar sind. Für eine eingehende Orientierung über psychoexperimentelle Untersuchungen am hirngeschädigten Menschen, die Untersuchungen über das Sprachzentrum sowie die Dominanz einer Rinden-Hemisphäre sei auf die einschlägige Literatur verwiesen (82).

Die Forderung von gleichwertigen psychologischen Testen vor und nach dem Hirntrauma erfüllt eine ausgedehnte Studie von WEINSTEIN und TEUBER (117) über die allgemeine Intelligenz nach Beschädigung des Hirnes. Diese Autoren machten Gebrauch von der in der amerikanischen Armee anläßlich der Rekrutenaushebung gebräuchlichen Intelligenztestung. Eine große Zahl von Soldaten wurden mit dem gleichen Intelligenztest 10 Jahre nach der im Krieg erfolgten Hirnverletzung wieder getestet und zugleich mit anderen Soldaten verglichen, bei denen als Kriegsfolge nur periphere Nerven, nicht aber das Hirn geschädigt worden war. Es zeigte sich, daß vor allem jene hirnlädierten Soldaten einen meßbaren und signifikanten Intelligenzverlust aufwiesen, bei denen die Kriegsverletzung den parietalen oder temporalen Cortex verletzt hatte. Weiter fanden die Autoren, daß bei allen Hirnverletzten als ganze Gruppe die Leistung im Intelligenztest seit der Rekrutenaushebung weniger zugenommen hatte als bei den Kontrollen. Dieser Hinweis auf die größere Wichtigkeit des parietalen und temporalen Cortex für die allgemeine Intelligenz als des frontalen Cortex steht aber mit vielen anderen Berichten nicht im Einklang. Andere Autoren wie z.B. ROSVOLD und MISHKIN (97) und HOYT, ELLIOTT und HEBB (36) testeten ebenfalls Soldaten, die vor dem Krieg mit dem gleichen Intelligenztest untersucht worden waren, und fanden einen größeren Intelligenzverlust nach Beschädigung des frontalen Cortex. Diese Gegenüberstellung vermag einen Hinweis auf die Schwierigkeiten des Problems zu geben. Erklärungen für die Verschiedenheit der Befunde mögen darin liegen, daß bei der Untersuchung von WEINSTEIN und TEUBER viel mehr Einzelindividuen untersucht wurden, und daß damit der Einfluß von Zufälligkeiten auf das Gesamtresultat besser vermieden werden konnte. Ferner war die Technik der Punktbewertung der Intelligenz in den verschiedenen Studien nicht einheitlich. Als allgemeiner Schluß dieser und eines Großteils von anderen Untersuchungen über die Intelligenz darf angenommen werden, daß ein echter Intelligenzverlust nach Cortexausschaltungen auftritt, daß aber die Frage nach der Gleichwertigkeit oder Nichtgleichwertigkeit verschiedener Cortexabschnitte für die allgemeine Intelligenz noch keineswegs entschieden ist. Sucht man in der Literatur nach Studien, die gezielt die Leistungsfähigkeit innerhalb einzelner Sinnesmodalitäten zu erfassen suchten, so trifft man auf die schon erwähnte Schwierigkeit, daß beim Menschen meistens Teste zur Anwendung gelangten, die gleichzeitig visuelle, akustische, verbale, Tastassoziationen usw. ohne klare Trennung durcheinander erfaßten. Daher ist es sehr schwierig, auf spezialisierte Leistungen einzelner Hirnfelder beim Menschen zu schließen. Es können daher in der Folge nur einige wenige Arbeiten erwähnt werden, auf die diese Kritik nicht zutrifft.

Eine Störung der Assoziationsfähigkeit für visuelle Eindrücke entsteht nach Milner (60) bei Ausschaltungen im Temporallappen. Die gleiche Autorin (62) zeigt in einer späteren Arbeit an einem viel umfangreicheren Kollektiv, daß für visuelle Unterscheidungen die Intaktheit des Temporallappens der nichtdominanten Hemisphäre wichtiger sei als jene der dominanten Hemisphäre, deren Ausschaltung vor allem die verbale Assoziationsfähigkeit beeinträchtige.

Störungen von Tastassoziationen sollen nach übereinstimmenden Berichten mehrerer Autoren vor allem verursacht werden durch Ausschaltungen im Parietallappen. Ghent et al. (28) fanden bei solchen Patienten, daß die Tastunterscheidung verschiedener Oberflächenqualitäten bei parietalen Ausschaltungen deutlich verlangsamt war. Weinstein et al. (118) kamen mit einem Test der Unterscheidungsfähigkeit für verschiedene Rauhheitsgrade zum gleichen Schluß. In einer anderen Arbeit (116) zeigte der gleiche Autor, daß solche Patienten mehr Fehler machen beim Schätzen von Gewichten, die ihnen in die Hand gelegt werden. Der Autor konnte auch zeigen, daß dieses Gewichtsschätzungsdefizit nur sichtbar wurde, wenn dem Patienten je ein Gewicht in beide Hände gelegt wird, und er sagen mußte, welches das schwerere war. Bekam der Patient zuerst das erste Gewicht in eine Hand, und dann später das andere Gewicht, so ließ sich der Patient von einem Kontrollpatienten nicht mit Sicherheit unterscheiden. Ferner zeigte es sich typischerweise, daß der einseitig hirngeschädigte Patient das Gewicht des Gegenstandes in der zur Verletzung kontralateralen Hand unterschätzt.

Diese wenigen Arbeiten zeigen, daß die Anwendung ähnlicher Teste wie im Tierversuch Resultate ergibt, die sich mit den experimentellen Ergebnissen beim Affen vergleichen lassen. Analoge Arbeiten über eine eventuelle Lokalisation der Unterscheidungsfähigkeit für andere Sinnesqualitäten wurden bis heute in dieser systematischen Weise nicht ausgeführt.

Als weitere Frage stellt sich, ob die Ausschaltung irgendeines Hirnteiles beim Menschen ein Versagen des „Sofortgedächtnisses" bewirke, wie es sich durch das Versagen in aufgeschobenen Aufgaben beim Macacen nach Verletzung des frontalen Cortex oder des Nucleus caudatus manifestiert. MILNER und PENFIELD (61) fanden ein solches Defizit nach beidseitiger tiefgehender Verletzung des Temporallappens. SCOVILLE und MILNER (104) fanden dann weiter, daß dieses Defizit davon abhänge, wie tief die Verletzung durch die temporale Hirnfläche in den Hippocampus eingedrungen ist. WALKER (114) fand ebenfalls, daß Verletzung des Hippocampus ein solches Defizit bewirke, doch ist es nach ihm nicht unbedingt notwendig, daß dazu der Hippocampus beidseitig ausgeschaltet werde. Diese Befunde stehen somit bis heute im Gegensatz zum Tierexperiment, wo das ähnliche Defizit nach Frontalhirnausschaltungen gefunden wurde. Doch liegen Berichte über Ausschaltung des Hippocampus und seine Folgen auf das „Sofortgedächtnis" beim Affen bis heute nicht vor. Zusammenfassend zu diesen Arbeiten darf gesagt werden, daß eine Übereinstimmung zwischen dem Tierbefund und dem Befund am Menschen dort nicht von der Hand zu weisen ist, wo in beiden Fällen eine analoge Untersuchungstechnik zur Anwendung gelangte. Dies geschah bis heute nur in wenigen Fällen, und nur wenige dieser Studien werden den strengen Kontrollbedingungen gerecht, die beim Experiment am Affen erfüllt werden konnten. Die Literatur über Sprachzentrum und Dominanz einer der beiden Hemisphären wird in diesem Zusammenhang nicht besprochen. Neuere Ergebnisse der Forschung auf diesem Gebiet haben PENFIELD und ROBERTS (82) in einer umfangreichen Übersicht zusammengestellt.

Zusammenfassung

Die Arbeit gibt eine Übersicht über die Wirkungen von Ausschaltungen im sekundären Cortex des Macacus-Affen auf dessen Leistungsfähigkeit in verschiedenen Lerntesten. Die wichtigsten Ergebnisse der besprochenen Untersuchungen sind in der Tabelle 1 zusammengestellt. Daraus geht hervor, daß für den sekundären oder Assoziationscortex des Macacus-Affen zwar eine deutlich erkennbare Aufteilung in funktionsspezifische Areale besteht, die jedoch in vielen Belangen noch einer eingehenderen Bearbeitung bedarf.

Die untere laterale und ventrale Fläche des Temporallappens ist essentiell für die visuelle Lernfähigkeit. Dieses Gebiet ist funktionell eng verbunden mit dem primären Sehcortex im Occipitallappen, dessen Ausschaltung im Gegensatz zu jener des temporalen Cortex nicht assoziative, sondern perzeptive Funktionsausfälle verursacht. Nach Ausschaltung des parietalen sekundären Cortex verlieren Affen die Fähigkeit, Tastunterscheidungen zu erlernen. Ein spezifisches Areal für das Erlernen auditorischer Unterscheidungen außerhalb des primären auditorischen Cortex konnte bisher nicht festgestellt werden. Ein Areal für das Erlernen von Geruchsunterscheidungen befindet sich im Temporallappen, ohne daß aber eine genauere Lokalisation, insbesondere eine

Tabelle 1. *Übersicht der experimentell festgestellten Wirkungen von Hirnausschaltung auf das Verhalten von Macacus-Affen* (Vergleiche mit Abb. 2 den Umfang der „Standardausschaltungen". Ohne besonderen Vermerk handelt es sich durchwegs um bilaterale Ausschaltungen)

Getestete Lernaufgabe	Fokalausschaltungen mit Verhaltensdefizit	Kontrollausschaltungen ohne Verhaltensdefizit	Charakteristika des Verhaltensdefizites
A. Verhaltensdefizite sensorischer Art			
Visuelle Unterscheidung: Erlernen, von zwei oder mehr simultan präsentierten visuellen Reizen den richtigen zu wählen	— temporale Standardausschaltung — einseitige temporale Standardausschaltung kombiniert mit kontralateraler occipitaler Standardausschaltung und Durchtrennung des posterioren Drittels des Corpus callosum	— nur einseitige temporale Standardausschaltung — superiorer temporaler Cortex — Durchtrennung des Corpus callosum — parietale Standardausschaltung — occipitale Standardausschaltung (verursacht nur geringfügiges Lerndefizit) — frontale Standardausschaltung — orbitaler frontaler Cortex — anteriorer insularer Cortex — periamygdaler Cortex — posteriorer Hippocampus — Pulvinar — Colliculi superiores	Lernen und Wiedererlernen sind postoperativ möglich, jedoch nur mit größerem Lernaufwand. Neulernen wird durch die Operation stärker beeinträchtigt als Wiedererlernen. Das Erlernen schwieriger Aufgaben ist stärker beeinträchtigt als jenes leichter Aufgaben. Eine Erholung der angelernten Funktion spontan mit der Zeit ist fraglich. Das Verhaltensdefizit ist nicht perzeptiver, sondern „integrativer" Natur. „Übertraining" vor der Operation verkleinert das postoperative Verhaltensdefizit
Visuelle Perception: Gesichtsfeld, Futtererkennung, Größenunterscheidung, Sehschärfe (Erkennen der Kreuzungen übereinandergelegter Fäden)	— occipitale Standardausschaltung	— temporale Standardausschaltung (verursacht nur geringfügiges Verhaltensdefizit) — frontale Standardausschaltung — frontotemporaler Cortex (Gyrus orbit. post., insularer und periamygdaler Cortex)	Das Verhaltensdefizit ist um so deutlicher, je schwieriger das „Problem" (kleine Differenzen im Größenunterscheidungstest, kompliziert hingelegte Fäden). Teilweise spontane Wiederholung der Funktion
Tastunterscheidung: Erlernen, von zwei simultan präsentierten Oberflächenqualitäten die richtige zu wählen	— parietale Standardausschaltung	— temporale Standardausschaltung — frontotemporaler Cortex (Gyrus orbitalis posterior, insularer und periamygdaler Cortex) — frontale Standardausschaltung	

Geruchsunterscheidung: Erlernen, von zwei simultan präsentierten Düften den richtigen zu wählen	— lateraler inklusive polarertemporaler Cortex	— temporale Standardausschaltung ?	Langsameres Erlernen und Wiedererlernen von Aufgaben nach der Operation
Geschmacksperzeption	— frontotemporaler Cortex (Gyrus orbitalis posterior, insularer und periamygdaler Cortex)	— temporale Standardausschaltung	Fressen von chininhaltigem Futter
Auditorische Unterscheidung: Je nach Qualität eines Tones Niederpressen des einen oder anderen von zwei Hebeln	— ? ?	— frontale Standardausschaltung — Nucleus caudatus	Für den Macacus-Affen sehr schwer zu erlernende Aufgabe

B. Verhaltensdefizite nicht sensorischer Art

Aufgeschobene Reaktion: Zwei Futterplätze, drei Testphasen: 1. Signalisierung des belohnten Futterplatzes mit positivem, des leeren Futterplatzes mit negativem Reiz. 2. Verdecken der Sicht auf die Situation während dem Aufschub. 3. Wahl des Tieres zwischen beiden, jetzt neutral zugedeckten Futterplätzen	— Gebiet beidseits und in der Tiefe des Sulcus principalis innerhalb der frontalen Standardausschaltung — Nucleus caudatus	— Cortex des Frontallappens außerhalb des Gebietes um den Sulcus principalis — temporale Standardausschaltung — parietale Standardausschaltung — occipitale Standardausschaltung — Nucleus dorsomedialis des Thalamus — motorischer und prämotorischer Cortex — prämotorische Augenfelder	Erlernen und Wiedererlernen sind postoperativ unmöglich. Keine spontane Wiedererholung. Defizit unabhängig von der Länge des Aufschubs im Test. Außer der klassischen Aufgabestellung sind postoperativ erlernbar: a) Korrekt-Reagieren oder Nicht-Reagieren bei nur einem Reiz nach Aufschub; b) wenn bei zwei Reizen dem Tier die Sicht auf beide Wahlorte während dem Aufschub belassen wird (Pseudoaufschub)
Hyperaktivität	— Ausschaltung in verschiedenen Gebieten des frontalen Cortex — Nucleus caudatus	— temporale Standardausschaltung — parietale Standardausschaltung — occipitale Standardausschaltung	Teilweise spontane Wiedererholung nach Operation. Typisches Im-Kreis Herumgehen

Tabelle 1 (Fortsetzung)

Getestete Lernaufgabe	Fokalausschaltungen mit Verhaltensdefizit	Kontrollausschaltungen ohne Verhaltensdefizit	Charakteristika des Verhaltensdefizites

C. „Sensorische" Verhaltungsdefizite unspezifischer Art

Getestete Lernaufgabe	Fokalausschaltungen mit Verhaltensdefizit	Kontrollausschaltungen ohne Verhaltensdefizit	Charakteristika des Verhaltensdefizites
Sukzessive Unterscheidung: Präsentation eines Einzelreizes (auditorisch, visuell). Reizqualität sagt, ob Futterplatz unter dem Einzelreiz leer oder voll ist	— dorsolateraler frontaler Cortex — Nucleus caudatus — temporale Standardausschaltung für visuelle Teste	— ?? — temporale Standardausschaltung für auditorische Teste	Verhaltensdefizit größer bei auditorischen als bei visuellen Aufgaben. Verzögertes Erlernen der negativen Reizbedeutung (= Hemmdefizit?)
Umlernen: Umlernen von positiver auf negative Reizbedeutung in simultanen Seriewahltesten	— dorsolateraler frontaler Cortex	— temporale Standardausschaltung	Perseveration der frontal operierten Tiere auf positiver Signalbedeutung
Schwache Strafmotivation: Schwache Elektroschockbestrafung bei audit. Test. Geringe physische Anstrengung bei Tastunterscheidung	— dorsolateraler frontaler Cortex	— ??	Postoperativ nur normales Erlernen, wenn Bestrafung oder physische Anstrengung zum Öffnen des Futterbehälters groß ist (=Verlust an Hemmfähigkeit)
Verschmelzungsfrequenz optischer Lichtblitze	— temporale Standardausschaltung — occipitale Standardausschaltung (Absinken der Verschmelzungsfrequenz)	— frontale Standardausschaltung (Ansteigen der Verschmelzungsfrequenz)	

Abgrenzung gegenüber dem visuellen assoziativen Areal, bisher bekannt wäre. Über die Lokalisation eines Feldes für Geschmacksunterscheidungen gibt es bis heute keine schlüssigen Studien.

Die totale Ausschaltung des frontalen sekundären Cortex bewirkt einen irreparablen Ausfall der Fähigkeit zur aufgeschobenen Reaktion. Die psychologische Bedeutung dieses Ausfalls ist bis heute noch unklar. Jedenfalls handelt es sich nicht um einen einfachen Gedächtnisverlust. Neben diesem Defizit bewirken Frontalausschaltungen noch andere Ausfälle vielfältiger Art, deren Interpretation auf Grund der heutigen Kenntnisse ebenfalls noch nicht gesichert ist. So geht die Fähigkeit, sensorische Unterscheidungen zu erlernen, für verschiedene Modalitäten verloren, aber nur, wenn der entsprechende Test in bestimmter Weise durchgeführt wird. Solche Ausfälle sind demnach nicht sinnes-spezifischer Art, wie dies bei den Ausschaltungen im posterioren sekundären Cortex (parietaler, temporaler und occipitaler sekundärer Cortex) der Fall war. Weitere Folgen von Frontalausschaltungen sind eine typische Hyperaktivität und in vielen Fällen eine Störung des normalen emotionellen Verhaltens.

Die Beziehung dieser Befunde zu der Humanpathologie liegt heute noch weitgehend im Dunklen, was vor allem daran zu liegen scheint, daß die gleiche Systematik der Untersuchungstechnik beim Menschen bis heute nur sehr wenig zur Anwendung gelangte.

Literatur

1. ALLEN, W. F.: Effect of ablating the frontal poles, hippocampi and occipito-parieto-temporal (excepting pyriform areas) lobes on positive and negative olfactory conditioned reflexes. Amer. J. Physiol. **128**, 754—771 (1940).
2. BAGSHAW, M. H., and K. H. PRIBRAM: Cortical organisation in gustation. J. Neurophysiol. **16**, 499—508 (1953).
3. BÄTTIG, K., H. E. ROSVOLD and M. MISHKIN: Comparison of the effects of frontal and caudate nucleus lesions on delayed response and alternation in monkeys. J. comp. physiol. **52**, 400—404 (1960).
4. — — — Comparison of the effects of frontal and candate lesions on dicrimination learning in monkeys. J. comp. physiol. Psychol. **55**, 458—463 (1962).
5. BIANCHI, L.: The mechanisms of the brain and the function of the frontal lobes. Trans. by J. H. MacDonald. New York: Wood & Co. 1922.
6. BLUM, J. S., K. L. CHOW and K. H. PRIBRAM: A behavioral analysis of the organisation of the parieto-temporo-preoccipital cortex. J. comp. Neurol. **93**, 53—100 (1950).
7. BLUM, A. R.: Effects of subtotal lesions of frontal granular cortex on delayed reaction in monkeys. Arch. Neurol. Psychiat. (Chic.) **67**, 375—386 (1952).
8. BONIN, G. v., and P. BAILEY: The neocortex of Macaca Mulatta. Urbana (Ill.): University Illinois Press 1947.
9. BRUSH, E. S., and M. MISHKIN: The relationship of object preference to learning set performance in brain operated monkeys. Paper presented at eastern psychological association meetings 1959.
10. BRUTKOWSKI, S., J. KONORSKI, W. LAWICKA, I. STEPIEN and L. STEPIEN: The effects of the removal of the frontal poles of the cerebral cortex on motor conditioned reflexes. Acta Biol. exp. (Lodz) **17**, 167—188 (1956).

11. Bruthowski, S.: The effect of prefrontal lobectomies on salivary conditioned reflexes in dogs. Acta Biol. exp. (Lodz) **17**, 327—337 (1957).

12. Butler, R. A., J. T. Diamond and W. D. Neff: Role of the auditory cortex in discrimination of changes of frequency. J. Neurophysiol. **20**, 108—120 (1957).

13. Chow, K. L.: Further studies on selective ablation of associative cortex in relation to visually mediated behavior. J. comp. physiol. Psychol. **48**, 229—237 (1952).

14. — Conditions influencing the recovery of visual discrimination habits in monkeys following temporal neocortical ablations. J. comp. physiol. Psychol. **45**, 430—437 (1952).

15. — effect of temporal neocortical ablation on visual discrimination learning sets in monkeys. J. comp. physiol. Psychol. **47**, 194—198 (1954).

16. — Lack of behavioral effects following destruction of some thalamic association nuclei. Arch. Neurol. Psychiat. (Chic.) **71**, 762—771 (1954).

17. —, and J. Orbach: Performance of visual discrimination presented tachistoscopically in monkeys with temporal neocortical ablations. J. comp. physiol. Psychol. **50**, 636—640 (1957).

18. Davis, G. D.: Effects of central excitant and depressant drugs on locomotor activity in the monkey. Amer. J. Physiol. **188**, 619—623 (1957).

19. Dean, W. H., and G. D. Davis: Behavior changes following caudate nucleus lesions in rhesus monkey. J. Neurophysiol. **22**, 524—538 (1959).

20. Diamond, J. T., and W. D. Neff: Ablation of temporal cortex and discrimination of auditory patterns. J. Neurophysiol. **20**, 300—315 (1957).

21. Ettlinger, G., and J. Wegener: Somaesthetic alternation, discrimination and orientation after frontal and parietal lesions in monkeys. Quart. J. exp. Psychol. **10**, 177—186 (1958).

22. — Visual discrimination following successive temporal ablations in monkeys. Brain **82**, 232—250 (1959).

23. Ferrier, D.: The Croonian lecture. Experiments on the brain of monkeys (second series). Philos. Trans. B **165**, 433—488 (1875).

24. Finan, J. L.: Delayed response with predelay reinforcement in monkeys after removal of the frontal lobes. Amer. J. Psychol. **55**, 202—214 (1942).

25. Forgays, D. G.: Reversible disturbances of function in man following cortical insult. J. comp. physiol. Psychol. **45**, 209—215 (1952).

26. French, G. M.: Locomotor effects of regional ablations of frontal cortex in rhesus monkeys. J. comp. physiol. Psychol. **52**, 18—24 (1959).

27. — A deficit associated with hypermotility in monkeys with lesions of the dorsolateral frontal granular cortex. J. comp. physiol. Psychol. **52**, 25—28 (1959).

28. Ghent, L., S. Weinstein, J. Semmes and H. L. Teuber: Effect of unilateral injury in man on learning of a tactual discrimination. J. comp. physiol. Psychol. **48**, 478—481 (1955).

29. Goldberg, J. M., J. T. Diamond and W. D. Neff: Auditory discrimination after ablation of temporal and insular cortex in cat. Fed. Proc. **16**, 47—48 (1957).

30. Hamuy, T. P., G. Santibanez, C. Gonzales and E. Vicencio: Changes in behavior and visual performance after selective ablation of the temporal cortex. J. comp. physiol. Psychol. **50**, 379—385 (1957).

31. Harlow, H. F.: Recovery of pattern discrimination in monkeys following unilateral occipital lobectomy. J. comp. physiol. Psychol. **27**, 467—489 (1939).

32. —, and J. Dagnon: Problem solution by monkeys following bilateral removal of the prefrontal areas: I. The discrimination and discrimination reversal problems. J. exp. Psychol. **32**, 351—356 (1943).

33. —, and P. H. Settlage: Effect of exstirpation of frontal areas upon learning performance of monkeys. Ass. Res. nerv. Dis. Proc. **27**, 446—459 (1948).

34. — R. T. Davis, P. H. Settlage and D. R. Meyer: Analysis of frontal and posterior association syndromes in brain damaged monkeys. J. comp. physiol. Psychol. **45**, 419—429 (1952).

35. HARMAN, P. J., M. H. C. TANKARD and F. A. METTLER: An experimental anatomical analysis of the topography and polarity of the caudate-neocortex interrelationship in the primate. Anat. Rec. 118, 307—308 (1954).
36. HOYT, R., H. ELLIOTT and D. O. HEBB: The intelligence of schizophrenic patients following lobotomy. Canad. Dept Veterans Affairs Bull. Treatment Services 6, 553—557 (1951).
37. JACOBSEN, C. F.: Functions of the frontal association areas in primates. Arch. Neurol. Psychiat. (Chic.) 33, 558—569 (1935).
38. — The functions of the frontal association areas in monkeys. Comp. Psychol. Monogr. 13, 3—60 (1936).
39. —, and J. H. ELDER: Studies of cerebral functions in primates: I. The functions of the frontal association areas in monkeys. Comp. Psychol. Monogr. 13, 61—65 (1936).
40. —, and G. M. HASLERUD: A note on the effect of motor and premotor area lesions on delayed response in monkeys. Comp. Psychol. Monogr. 13, 66—68 (1936).
41. JASPER, H. H., C. AJMONE MARSAN and J. STOLL: Corticofugal projections to the brain stem. Arch. Neurol. Psychiat. (Chic.) 67, 155—166 (1952).
42. KAADA, B. R.: Somato-motor, autonomic and electrocorticographic stimulation of "rhinencephalic" and other structures in primats cat and dog. Acta physiol. scand. 24, Suppl. 83 (1951).
43. KENNARD, M. A., S. SPENCER and G. FOUNTAIN: Hyperactivity in monkeys following lesions of the frontal lobes. J. Neurophysiol. 4, 512—524 (1941).
44. KLÜVER, H.: Certain effects of lesions of the occipital lobes in macaques. J. Psychol. (Provincetown) 4, 383—401 (1937).
45. —, and P. C. BUCY: Preliminary analysis of functions of the temporal lobes in monkeys. Arch. Neurol. Psychiat. (Chic.) 42, 979—1000 (1939).
46. KÖHLER, W., and R. HELD: The cortical correlate of pattern vision. Science 110, 414—419 (1949).
47. LASHLEY, K. S.: The mechanism of vision. XVIII. Effects of destroying the visual "associative areas" of the monkey. Genet. Psychol. Monogr. 37, 107—166 (1948).
48. — K. L. CHOW and J. SEMMES: An examination of the electrical field theory of cerebral integration. Psychol. Rev. 58, 123—136 (1951).
49. LAWICKA, W.: The effect of the prefrontal lobectomy on the vocal conditioned reflexes in dogs. Acta Biol. exp. (Lodz) 17, 317—325 (1957).
50. LICHTENSTEIN, P. E.: Studies of anxiety: II. The effect of lobotomy on a feeding inhibition in dogs. J. comp. physiol. Psychol. 43, 419—427 (1950).
51. LIVINGSTONE, R. B., J. F. FULTON, J. M. R. DELGADO, E. SACHS jr., S. J. BRENDLER and G. DAVIS: Stimulation and regional ablation of orbital surface of frontal lobe. Ass. Res. nerv. Dis. Proc. 27, 405—420 (1948).
52. MALMO, R. B.: Interference factors in delayed response in monkeys after removal of frontal lobes. J. Neurophysiol. 5, 295—308 (1942).
53. METTLER, F. A., and C. C. METTLER: The effects of strial injury. Brain 65, 242—253 (1942).
54. — Effects of bilateral simultaneous subcortical lesions in the primate. J. Neuropath. exp. Neurol. 4, 99—122 (1945).
55. — C. HOVDE and H. GRUNDFEST: Electrophysiologic phaenomena evoked by stimulation of caudate nucleus. Fed. Proc. 11, 107 (1952).
56. MEYER, R. D., H. F. HARLOW and P. H. SETTLAGE: A survey of delayed response performance by normal and brain damaged monkeys. J. comp. physiol. Psychol. 44, 17—25 (1951).
57. —, and P. H. SETTLAGE: Analysis of simple searching behavior in the frontal monkey. J. comp. physiol. Psychol. 51, 408—410 (1958).
58. MIGLER, B.: The effect of lesions to the caudate nuclei and corpus callosum on delayed alternation in the monkey. Master-Thesis, University of Pittsburgh (Pa.) 1958.

59. Miles, J. C., and H. E. Rosvold: The effects of prefrontal lobotomy in rhesus monkeys on delayed response performance motivated by pain shock. J. comp. physiol. Psychol. **49**, 286—292 (1956).

60. Milner, B.: Intellectual functions of the temporal lobes. Psychol. Bull. **51**, 42—62 (1954).

61. —, and W. Penfield: The effect of hippocampal lesions on "recent memory". Trans. Amer. neurol. Ass. **80**, 42—48 (1955).

62. — Psychological defects produced by temporal lobe excision. Res. Publ. Ass. nerv. ment. Dis. **36**, 244—257 (1956).

63. Mishkin, M., H. E. Rosvold and K. H. Pribram: Effects of nembutal in baboons with frontal lesions. J. Neurophysiol. **16**, 155—159 (1953).

64. —, and K. H. Pribram: Visual discrimination performance following partial ablations of the temporal lobe. I. Ventral vs. lateral. J. comp. physiol. Psychol. **47**, 14—20 (1954).

65. — Visual discrimination performance following partial ablations of the temporal lobe. II. Ventral vs. Hippocampus. J. comp. physiol. Psychol. **47**, 187—193 (1954).

66. —, and M. Hall: Discrimination along a size continuum following partial ablations of the temporal lobe. J. comp. physiol. Psychol. **48**, 97—101 (1955).

67. —, and K. H. Pribram: Analysis of the effects of frontal lesions in monkeys. I. Variations of delayed alternation. J. comp. physiol. Psychol. **48**, 492—495 (1955).

68. — — Analysis of the effects of frontal lesions in monkeys: II. Variations of delayed response. J. comp. physiol. Psychol. **49**, 36—40 (1956).

69. — Effects of small frontal lesions on delayed alternation in monkeys. J. Neurophysiol. **20**, 615—622 (1957).

70. — Visual discrimination impairment after cutting cortical connections between the inferotemporal and striate areas in monkeys. Amer. Psychologist **13**, 414 (1958).

71. —, and L. Weiskrantz: Effects of delaying reward on visual discrimination performance in monkeys with frontal lesions. J. comp. physiol. Psychol. **51**, 276—281 (1958).

72. — — Effects of cortical lesions in monkeys on critical flicker frequency. J. comp. physiol. Psychol. **52**, 660—667 (1959).

73. Myers, R. A.: Interocular transfer of pattern discrimination in cats following section of crossed optic fibers. J. comp. physiol. Psychol. **48**, 470—473 (1955).

74. Myers, R. E.: Function of corpus callosum in interocular transfer. Brain **79**, 358—363 (1956).

75. — Localisation of function within the corpus callosum visual gnostic transfer. Anat. Rec. **124**, 339—340 (1956).

76. Neff, W. D., J. F. Fisher, I. T. Diamond and M. Yala: Role of auditory cortex in discrimination requiring localisation of sound in space. J. Neurophysiol. **19**, 500—512 (1956).

77. — Behavioral studies of auditory discrimination. Ann. Otol. (St. Louis) **66**, 506—513 (1957).

78. Niemer, W. T., and J. Jimenez Castellanos: Cortico-thalamic connections in the cat as revealed by physiological neuronography. J. comp. Neurol. **93**, 101—123 (1950).

79. Orbach, J.: Immediate and chronic disturbances on the delayed response following transsection of frontal granular cortex in the monkey. J. comp. physiol. Psychol. **49**, 46—51 (1956).

80. Orbach, H., and R. L. Fantz: Differential effects of temporal neocortical resections on overtrained and non-overtrained visual habits in monkeys. J. comp. physiol. Psychol. **51**, 126—129 (1958).

81. Orbach, J., and K. L. Chow: Differential effects of resections of somatic areas I and II in monkeys. J. Neurophysiol. **22**, 195—203 (1959).

82. Penfield, W., and L. Roberts: Speech and brain mechanisms. Princeton: Princeton University Press 1959.
83. Pasik, P., T. Pasik, W. S. Battersby and M. B. Bender: Effects of serial temporal and parietal lesions upon discriminative learning in macaques. Fed. Proc. **15**, 142 (1956).
84. Peters, R. H., H. E. Rosvold and A. F. Mirsky: The effect of thalamic lesions upon delayed response type tests in the rhesus monkey. J. comp. physiol. Psychol. **49**, 111—116 (1956).
85. Pribram, H. B., and J. Barry: Further behavioral analysis of parieto-temporo-preoccipital cortex. J. Neurophysiol. **19**, 99—106 (1956).
86. Pribram, K. H.: Some physical and pharmacological factors affecting delayed response performance of baboons following frontal lobotomy. J. Neurophysiol. **13**, 373—382 (1950).
87. — M. Mishkin, H. E. Rosvold and S. J. Kaplan: Effects on delayed response performance of lesions of dorsolateral and ventromedial frontal cortex of baboons. J. comp. physiol. Psychol. **45**, 565—575 (1952).
88. —, and M. Bagshaw: Further analysis of temporal lobe syndrome utilizing fronto-temporal ablations. J. comp. Neurol. **99**, 347—375 (1953).
89. —, and L. Kruger: Functions of the olfactory brain. Ann. N.Y. Acad. Sci. **58**, 109—138 (1954).
90. — Lesions of the "frontal eye fields" and delayed response of baboons. J. Neurophysiol. **18**, 105—112 (1955).
91. —, and M. Mishkin: Simultaneous and successive visual discrimination by monkeys with inferotemporal lesions. J. comp. physiol. Psychol. **48**, 198—202 (1955).
92. — — Analysis of the effects of frontal lesions in monkeys: III. Object alternation. J. comp. physiol. Psychol. **49**, 41—45 (1956).
93. —, and L. Weiskrantz: A comparison of the effects of medial and lateral cerebral resections on conditioned avoidance behavior in monkeys. J. comp. physiol. Psychol. **50**, 74—80 (1957).
94. Richter, C. P., and M. Hines: Increased spontaneous activity produced in monkeys by brain lesions. Brain **61**, 1—16 (1938).
95. Riopelle, A. J., and H. W. Ades: Discrimination learning following deep tempora lesions. Amer. Psychologist **6**, 261 (1951) (Abstract).
96. —, and G. A. Churukian: The effect of varying the intertrial interval in discrimination learning by normal and brain operated monkeys. J. comp. physiol. Psychol. **51**, 119—125 (1958).
97. Rosvold, H. E., and M. Mishkin: Evaluation of the effects of prefrontal lobotomy on intelligence. Canad. J. Psychol. **3**, 122—126 (1950).
98. —, and J. M. R. Delgado: The effect on delayed alternation test performance of stimulating or destroying electrically structures within the frontal lobes of the monkey's brain. J. comp. physiol. Psychol. **49**, 365—372 (1956).
99. — M. Mishkin and M. K. Szwarcbart: Effects of subcortical lesions in monkeys on visual discrimination and single alternation performance. J. comp. physiol. Psychol. **51**, 437—444 (1958).
100. —, and M. Mishkin: Non sensory effects of frontal lesions on discrimination learning and performance in J. F. Delaresnaye (Ed.), Brain mechanisms and learning. Oxford: Blackwell (1961)
101. — Personal communication 1961.
102. Ruch, T. C., and H. A. Shenkin: The relation of area 13 of the orbital surface of the frontal lobes to hyperactivity and hyperphagia in monkeys. J. Neurophysiol. **6**, 349—360 (1943).
103. Santibanez, G., and T. P. Hamuy: Olfactory discrimination deficits in monkeys with temporal lobe ablations. J. comp. physiol. Psychol. **50**, 472—474 (1957).
104. Scoville, W. B., and B. Milner: Loss of recent memory after bilateral hippocampal lesions. J. Neurol. Neurosurg. Psychiat. **20**, 11—21 (1957).

105. Settlage, P. H.: The effect of occipital lesions on visually guided behavior in the monkey. J. comp. Psychol. 27, 93—131 (1939).
106. Spence, K. W., and J. F. Fulton: The effects of occipital lobectomy on vision in chimpanzee. Brain 59, 35—50 (1936).
107. Sperry, R. W., N. Miner and R. W. Meyers: Visual pattern perception following subpial slicing and tantalum implantations in the visual cortex. J. comp. physiol. Psychol. 48, 50—58 (1955).
108. Sperry, R. W., J. S. Stamm and N. Miner: Relearning tests for interocular transfer following division of optic chiasma and corpus callosum in cats. J. comp. physiol. Psychol. 49, 529—533 (1956).
109. Stanley, W. C., and J. Jaynes: The function of the frontal cortex. Psychol. Rev. 56, 18—32 (1949).
110. Talbot, S. A., and W. H. Marshall: Physiological studies on neural mechanisms of visual localisation and discrimination. Amer. J. Ophthal. 24, 1255—1264 (1941).
111. Thomas, J., and P. A. Stewart: The effect on visual perception of stimulating the brain with polarizing currents. Amer. J. Psychol. 70, 528—540 (1957).
112. Wade, M.: The effect of the sedatives upon delayed response in monkeys after removal of prefrontal lobes. J. Neurophysiol. 10, 57—62 (1947).
113. — Behavioral effects of prefrontal lobectomy, lobotomy and circumsection in the monkey. J. comp. Neurol. 96, 179—207 (1952).
114. Walker, A. E.: "Recent memory" impairment in unilateral temporal lesions. Arch. Neurol. Psychiat. (Chic.) 78, 543—552 (1957).
115. Waterhouse, J. K.: Effects of prefrontal lobotomy on conditioned fear and food responses in monkeys. J. comp. physiol. Psychol. 50, 81—88 (1957).
116. Weinstein, S.: Weight judgement in somaesthesis after penetrating injury to the brain. J. comp. physiol. Psychol. 47, 31—35 (1954).
117. —, and H. L. Teuber: Effects of penetrating brain injury on intelligence test scores. Science 125, 1036—1037 (1957).
118. — J. Semmes, L. Ghent and H. L. Teuber: Roughness discrimination after penetrating brain injury in man: Analysis according to locus of lesions. J. comp. physiol. Psychol. 51, 269—275 (1958).
119. Weiskrantz, L.: Behavioral changes associated with ablation of the amygdaloid complex in monkey. J. comp. physiol. Psychol. 49, 381—391 (1956).
120. —, and M. Mishkin: Effects of temporal and frontal cortical lesions on auditory discrimination in monkeys. Brain 81, 406—414 (1958).
121. Whitlock, D. G., and W. J. H. Nauta: Subcortical projections from the temporal neocortex in macaca mulatta. J. comp. Neurol. 106, 183—212 (1956).
122. Wilson, M.: Effects of circumscribed cortical lesions upon somaesthetic and visual discrimination in the monkey. J. comp. physiol. Psychol. 50, 630—635 (1957).
123. Wilson, A. W., and M. Mishkin: Comparison of the effects of interferotemporal and lateral occipital lesions on visually guided behavior in monkey. J. comp. physiol. Psychol. 52, 10—17 (1959).

Namenverzeichnis

Die in Klammern stehenden Ziffern beziehen sich auf die Nummern der Zitate innerhalb des laufenden Textes und der Literatur.

Die gewöhnlich gesetzten Ziffern weisen auf die entsprechende Stelle im Text und die *kursiven* Seitenzahlen auf das Literaturverzeichnis hin.

Sachverzeichnis

barrier phenomena and metabolism of nervous tissue 39
basement membrane 34, 35, 47
baroreceptors, see systemic arterial baroreceptors and pulmonary arterial baroreceptors
bedingte Reflexe 158
bedingte Speichelreflexe nach Frontalausschaltung 190
Bedingungswahl 162, 168
B-elevation of the compound action potential of the vagus 88
Blindheit nach Parietalausschaltung 177
blocking temperatures of afferent fibres from atrial receptors 90, 92
— — — — systemic arterial baroreceptors 90, 92
— — — — pulmonary stretch receptors 90, 91
— — — — rapidly adapting tracheobronchial receptors 90, 91
blood-brain barrier 23f., 25, 29, 33f., 46
— — to adrenaline 62
— — and air embolism 61
— —, allergic response 61
— — to amino acids 39
— —, breakdown of 31, 35, 37
— — — — to fluorescein 32
— — to Ca 62
— —, concentration gradient across 40
— — to Diamox 49
— —, after electroshocks 62
— — and exposure of brain 62
— —, functional studies 36f.
— — to histamine 62
— — and hypothermia 62
— —, hypoxia 62
— — to iodide 37
— — and local cold injury 61
— —, metabolic aspects 39
— — to methionine 38
— —, morphological correlate of 33f.
— — to ^{24}Na 37
— —, ontogeny 38
— — to PA-H 62
— —, penetration of determined by lipid-solubility 23, 24
— — to $^{32}PO_4$ 39
— — to proline 62
— — to Sr 62
— — to sulphanilic acid 62
— — to thiocyanate 37
— — to triiodothyronine 62
— — to urea 41

blood-cerebrospinal fluid barrier 21f.
— — — — to Diodrast 42
— — — — as a function of choroidal epithelium 35
— — — — to urea 41
blood volume, thoracic, and atrial activity 116
brain, area of the surface 38
—, atrophy and rise in β-globulin 54
—, extracellular space of 24f.
—, exposure of and blood-brain barrier 62
—, perfused and utilization of glucose 40
—, shrinkage following administration of urea 41
—, sodium-space of 32
—, synthesis of urea 41
—, uptake of $^{32}PO_4$ 38
—, venous effluent from 24
—, wateruptake and electrolyte pumps 32
—, watercontent and osmotic pressure in the blood 31
—, damage by cold 36
—, — — ultrasonic irradiation 37
—, oedema, complex 32
—, — due to compression 32
—, — due to heat-damage 32
—, —, interstitial 30, 32, 33
—, —, intracellular 33
—, — due to local cooling of the pia 32
—, —, simple 32
— tissue, absorption of substances from cerebrospinal fluid 44
— —, diffusional relations to cerebrospinal fluid 42
— —, impedance of 29
— —, metabolism of and barrier phenomena 39
— —, pressure, effects of drugs on 51
— — as source of cerebrospinal fluid proteins 52, 54, 55
— — — — neuraminic acid 58
bronchi, extrapulmonary, location of bronchial receptors 101
—, intrapulmonary, location of pulmonary stretch receptors 100
bronchial receptors, slowly adapting 95, 100f, 106
— —, sensitization of 101
— tone and acetylcholine 100
— —, adaptation of pulmonary stretch receptors 98
bronchioles, respiratory, location of deflation receptors 110

ERGEBNISSE DER PHYSIOLOGIE
BIOLOGISCHEN CHEMIE UND
EXPERIMENTELLEN PHARMAKOLOGIE

HERAUSGEGEBEN VON

K. KRAMER
GÖTTINGEN

O. KRAYER
BOSTON

E. LEHNARTZ
MÜNSTER / WESTF.

A. v. MURALT
BERN

H. H. WEBER
HEIDELBERG

BAND 52

SONDERDRUCK

H. H. DALE

OTTO LOEWI †

WITH 1 PORTRAIT

NICHT IM HANDEL

SPRINGER-VERLAG
BERLIN · GÖTTINGEN · HEIDELBERG
1963

Inhaltsverzeichnis
Band 52

ERGEBNISSE DER PHYSIOLOGIE
BIOLOGISCHEN CHEMIE UND EXPERIMENTELLEN PHARMAKOLOGIE

HERAUSGEGEBEN VON

K. KRAMER
GÖTTINGEN

O. KRAYER
BOSTON

E. LEHNARTZ
MÜNSTER / WESTF.

A. v. MURALT
BERN

H. H. WEBER
HEIDELBERG

BAND 52

SONDERDRUCK

HUGH DAVSON

THE CEREBROSPINAL FLUID

WITH 8 FIGURES

NICHT IM HANDEL

SPRINGER-VERLAG
BERLIN · GÖTTINGEN · HEIDELBERG
1963

Inhaltsverzeichnis
Band 52

ERGEBNISSE DER PHYSIOLOGIE
BIOLOGISCHEN CHEMIE UND
EXPERIMENTELLEN PHARMAKOLOGIE

HERAUSGEGEBEN VON

K. KRAMER
GÖTTINGEN

O. KRAYER
BOSTON

E. LEHNARTZ
MÜNSTER / WESTF.

A. v. MURALT
BERN

H. H. WEBER
HEIDELBERG

BAND 52

SONDERDRUCK

A. S. PAINTAL

VAGAL AFFERENT FIBRES

WITH 36 FIGURES

NICHT IM HANDEL

SPRINGER-VERLAG
BERLIN · GÖTTINGEN · HEIDELBERG
1963

Inhaltsverzeichnis
Band 52

ERGEBNISSE DER PHYSIOLOGIE
BIOLOGISCHEN CHEMIE UND
EXPERIMENTELLEN PHARMAKOLOGIE

HERAUSGEGEBEN VON

K. KRAMER
GÖTTINGEN

O. KRAYER
BOSTON

E. LEHNARTZ
MÜNSTER / WESTF.

A. v. MURALT
BERN

H. H. WEBER
HEIDELBERG

BAND 52

SONDERDRUCK

K. BÄTTIG UND H. E. ROSVOLD

PSYCHOPHYSIOLOGISCHE LEISTUNGSFÄHIGKEIT DES MACACUSAFFEN NACH CORTEXAUSSCHALTUNGEN

MIT 5 ABBILDUNGEN

NICHT IM HANDEL

SPRINGER-VERLAG

BERLIN · GÖTTINGEN · HEIDELBERG

1963

Inhaltsverzeichnis
Band 52

Ergebnisse der Physiologie
biologischen Chemie und experimentellen Pharmakologie

Herausgegeben von K. Kramer, Göttingen, O. Krayer, Boston, E. Lehnartz, Münster i. W., A. v. Muralt, Bern, H. H. Weber, Heidelberg.

Zuletzt erschienene Bände

Fünfzigster Band: Mit 57 Abbildungen. IV, 564 Seiten (davon 321 Seiten in englischer und 37 Seiten in französischer Sprache) Gr.-8°. 1959. Steif geheftet DM 118,—

Inhaltsübersicht

Thrombocytenfaktoren. Von E. F. Lüscher, Bern (Schweiz). — The biosynthesis of the purines. By St. C. Hartman, Boston/Mass. (USA), and J. M. Buchanan, Cambridge/Mass. (USA). — Récentes acquisitions sur la nature et le métabolisme des hormones thyroïdiennes. Par J. Roche et R. Michel, Paris (France). — Ionic movements in cell membranes in relation to the activity of the nervous system. By H. H. Ussing, Copenhagen (Denmark). — Mucosaccharides and glycoproteins. Chemistry and physiopathology. By Z. Stary, Warren, Pa. (USA). — Renin and hypertensin. By W. S. Peart, London (Great Britain). — Das Nierenmark. Struktur, Stoffwechsel und Funktion. Von K. J. Ullrich, Göttingen. — Namen- und Sachverzeichnis.

Einundfünfzigster Band: Mit 106 Abbildungen und 3 Porträts. VI, 469 Seiten (davon 178 Seiten in englischer Sprache) Gr.-8°. 1961. Steif geheftet DM 98,—

Inhaltsübersicht

Nachruf auf K. U. Linderstrøm-Lang †. Von Professor Dr. F. Duspiva, Heidelberg. — Nachruf auf Otto F. Ranke †. Von Professor Dr. W. D. Keidel, Erlangen. — Nachruf auf Ulrich Ebbecke †. Von Professor Dr. Hans Schaefer, Heidelberg. — Pyridine nucleotide dependent metallodehydrogenases. By Dr. Bert L. Vallee and Dr. Frederic L. Hoch, Boston, Mass. (USA). — Über die verschiedenen molekularen Mechanismen der Bewegungen von Zellen. Von Privatdozent Dr. Hartmut Hoffmann-Berling, Heidelberg. — Elektrophysiologie der Herzmuskelfaser. Von Professor Dr. Wolfgang Trautwein, Salt Lake City, Utah (USA). — Extrakardiale Digitaliswirkungen. Von Professor Dr. L. Lendle, Göttingen, und Professor Dr. H. Mercker, Stolberg/Rhld. — The mechanism of synaptic transmission. By Professor Dr. John C. Eccles, Canberra (Australia). — Namen- und Sachverzeichnis.

SPRINGER-VERLAG · BERLIN · GÖTTINGEN · HEIDELBERG